Famous Names in En...

To my Daughters Helen and Sarah

Famous Names in Engineering

James Carvill BSc(Hons)
Senior Lecturer in Mechanical Engineering,
Newcastle-upon-Tyne Polytechnic

Butterworths
London Boston
Sydney   Wellington   Durban   Toronto

The Butterworth Group

United Kingdom
Butterworth & Co. (Publishers) Ltd
London: 88 Kingsway, WC2B 6AB

Australia
Butterworths Pty Ltd
Sydney: 586 Pacific Highway, Chatswood, NSW 2067
Also at Melbourne, Brisbane, Adelaide and Perth

Canada
Butterworth & Co. (Canada) Ltd
Toronto: 2265 Midland Avenue, Scarborough, Ontario, M1P 4S1

New Zealand
Butterworths of New Zealand Ltd
Wellington: 31-35 Cumberland Place, CPO Box 472

South Africa
Butterworth & Co. (South Africa) (Pty) Ltd
Durban: 152-154 Gale Street

USA
Butterworth Publishers, Inc.
Boston: 10 Tower Office Park, Woburn, Mass. 01801

First published 1981

© Butterworth & Co. (Publishers) Ltd, 1981

All rights reserved. No part of this publication may be reproduced or transmitted in any form or by any means, including photocopying and recording, without the written permission of the copyright holder, application for which should be addressed to the Publishers. Such written permission must also be obtained before any part of this publication is stored in a retrieval system of any nature.

This book is sold subject to the Standard Conditions of Sale of Net Books and may not be re-sold in the UK below the net price given by the Publishers in their current price list.

British Library Cataloguing in Publication Data

Carvill, James
  Famous names in engineering.
  1. Scientists — Biography 2. Engineers — Biography
  I. Title
  509'.2'2   Q141   80-41705

  ISBN 0-408-00540-8 (limp)
  ISBN 0-408-00539-4 (cased)

Typeset by Tunbridge Wells Typesetting Services
Printed in England by Redwood Burn Ltd, Trowbridge and Esher

Preface

In lectures to students taking advanced courses in Engineering, the names of many famous mathematicians, physicists, and engineers are mentioned frequently in connection with theories, laws, formulae and inventions. Unfortunately however, in many cases neither the student nor the lecturer knows much more than the surname of the person. Where and when was he born? Who were his contemporaries and how was his work affected by the events of the times? What did he achieve in other fields and what interesting anecdotes are associated with him? These and many other questions can only be answered by reference to lengthy biographies, if they exist, or to biographical dictionaries comprising many volumes. Even there the information is often scanty.

The main purpose of this book is to bring to life as it were, the person behind the formula or law. It is hoped that the student's interest will therefore be stimulated by a knowledge of the lives and works of these pioneers of technology. The lecturer, professional engineer and the layman interested in engineering history should find the book a handy and inexpensive reference work.

Famous Names In Engineering gives brief and, it is hoped, interesting and informative biographies of 83 famous people whose laws, theories, and inventions form the basis of any advanced course in Engineering. They are in alphabetical order and in each case a portrait is included together with a relevant formula, diagram or illustration. The book contains a Chronological Table and also a Table of Subjects with lists of names under the appropriate subject headings. For those readers who desire further information, the entries are followed by a list of authoritative references.

Acknowledgement and thanks are due to the many organisations whose valuable assistance in supplying information helped so materially in the compiling of the book. Their names are included in the 'Further Reading' section.

I should like to express my thanks to colleagues at Newcastle-upon-Tyne Polytechnic for reading the manuscript and making constructive suggestions, to Ian Winship and Jeremy Atkinson of the Polytechnic Library for their invaluable assistance, and finally to Mark Cook who helped with the processing of the illustrations.

J.C.

Contents

Famous Names, by Subject *viii*

Famous Names, Chronologically *x*

Ampère, André Marie *1*

Archimedes of Syracuse *1*

Armstrong, Sir William George *2*

Bernoulli, Daniel *3*

Bessemer, Henry *5*

Bourdon, Eugéne *6*

Boyle, Robert *7*

Bramah, Joseph *8*

Callendar, Hugh Longbourne *9*

Carnot, Nicolas Léonard Sadi *10*

Castigliano, Carlo Alberto *11*

Celsius, Anders *12*

Charles, Jacques Alexandre César *13*

Clapeyron, Bénoit Pierre Émile *14*

Clausius, Rudolf Julius Emmanuel *15*

Coriolis, Gaspard Gustav de *15*

Coulomb, Charles Augustin de *16*

Curtis, Charles Gordon *18*

D'Alembert, Jean le Rond *18*

Darcy, Henri Philibert Gaspard *19*

De Laval, Carl Gustav Patrik *20*

Descartes, René *21*

Diesel, Rudolf *22*

Ericsson, John *23*

Euler, Leonhard *24*

Fahrenheit, Gabriel Daniel *25*

Fanning, John Thomas *26*

Faraday, Michael *27*

Fourier, Baron Jean Baptiste Joseph *28*

Fourneyron, Benoit *29*

Francis, James Bicheno *30*

Froude, William *31*

Galvani, Luigi *32*

Gauss, Karl Friedrich *33*

Henry, Joseph *34*

Hero of Alexandria *35*

Hertz, Heinrich Rudolf *36*

Hooke, Robert *36*

Joule, James Prescott *38*

Kármán, Théodore von *39*

Kelvin, Lord (William Thomson) *40*

Kirchhoff, Gustav Robert *41*

Lagrange, Joseph Louis *41*

Lamé, Gabriel *42*

Laplace, Pierre Simon *43*

Lenoir, Jean Joseph Étienne *44*

Leonardo da Vinci *45*

Mach, Ernst *46*

Magnus, Heinrich Gustav *47*

Maxwell, James Clerk *48*

Mohr, Christian Otto *49*

Mollier, Richard *50*

Newton, Sir Isaac *51*

Nusselt, Ernst Kraft Wilhelm *52*

Ohm, Georg Simon *53*

Otto, Niklaus August *54*

Parsons, Sir Charles Algernon *55*

Pascal, Blaise *56*

Pelton, Lester Allen *57*

Pitot, Henri *58*

Planck, Max Carl Ernst Ludwig *59*

Poiseuille, Jean Louis Marie *60*

Poisson, Simeon Denis *61*

Prandtl, Ludwig *61*

Pythagoras of Samos *62*

Rankine, William John MacQuorn *63*

Rayleigh, Lord (John William Strutt) *64*

Reynolds, Osborne *65*

Rumford, Count (Benjamin Thompson) *66*

Siemens, Ernst Werner von *67*

Stephenson, George *69*

Stirling, Robert *70*

Stokes, Sir George Gabriel *71*

Tesla, Nikola *72*

Timoshenko, Stepan Prokf'yevich *73*
Toricelli, Evangelista *75*
Trevethick, Richard *76*
Venturi, Giovanni Battista *77*
Volta, Count Alessandro *78*

Watt, James *79*
Weber, Wilhelm Eduard *80*
Whitworth, Sir Joseph *81*
Young, Thomas *82*
Further Reading *83*

Famous Names, by Subject

Aerodynamics
Charles
Kármán
Leonardo
Mach
Magnus
Prandtl

Astronomy
Celsius
Gauss
Hooke
Laplace
Newton
Parsons
Pythagoras

Ballistics
Archimedes
Armstrong
Bessemer
Ericsson
Lamé
Count Rumford
Whitworth

Chemistry
Ampère
Boyle
Kirchhoff
Magnus

Dimensional Analysis
Froude
Mach
Nusselt
Prandtl
Rayleigh
Reynolds

Electricity and Magnetism
Ampère
Coulomb
Faraday
Fourier
Galvani
Gauss
Henry
Hertz
Kelvin
Kirchhoff
Maxwell
Ohm
Poisson
Siemens
Tesla
Volta
Weber

Fluid Mechanics
Archimedes
Armstrong
Bernoulli
Bourdon
Bramah
D'Alembert
Darcy
Ericsson
Euler
Fanning
Fourneyron
Francis
Froude
Herschel
Hero
Kármán
Leonardo
Mach
Magnus
Newton
Nusselt
Pelton
Pitot
Poiseuille
Rayleigh
Reynolds
Stokes
Venturi

Fluid Statics
Archimedes
Pascal
Toricelli

Heat Engines
Callendar
Carnot
Clausius
Curtis de Laval
Diesel
Ericsson
Hero
Joule
Lenoir
Mollier
Otto
Parsons
Rankine
Stephenson
Stirling
Trevethick
Watt

Heat Transfer
Fourier
Nusselt
Planck
Prandtl
Reynolds

Hydraulic Power
Armstrong
Bramah
Fourneyron
Francis
Hero
Pelton

Industrialists
Armstrong
Bessemer
Bramah
De Laval
Diesel
Ericsson
Parsons
Stephenson
Tesla
Trevethick
Watt
Whitworth

Mathematics
Ampère
Archimedes
Bernoulli
Descartes
Euler
Fourier
Gauss
Henry
Hero
Kelvin
Lagrange
Lamé
Laplace
Maxwell
Newton
Poisson
Pythagoras
Rayleigh
Stokes

Measurements and Instrumentation
Bourdon
Callendar
Celsius
Charles

Coulomb
Fahrenheit
Froude
Herschel
Hooke
Kelvin
Pitot
Poiseuille
Rayleigh
Toricelli
Weber
Whitworth

Mechanics
Archimedes
Coriolis
Coulomb
D'Alembert
Fourier
Hooke
Lagrange
Leonardo
Newton
Rayleigh
Reynolds
Timoshenko

Metal Forming and Machining
Bramah
Whitworth

Metallurgy
Bessemer
Siemens

Philosophy
Boyle
D'Alembert
Descartes
Mach
Pythagoras

Physics
Ampère
Descartes
Faraday
Gauss
Maxwell
Newton
Ohm
Young

Ship Hydrodynamics
Armstrong
Ericsson
Froude
Parsons

Strengths of Materials and Elasticity
Castigliano
Clapeyron
Euler
Hertz
Hooke
Kirchhoff
Lamé
Maxwell
Mohr
Poisson
Prandtl
Rankine

Timoshenko
Young

Thermodynamics
Bourdon
Boyle
Callendar
Carnot
Charles
Clapeyron
Clausius
De Laval
Diesel
Ericsson
Hero
Joule
Kármán
Kelvin
Lenoir
Maxwell
Mollier
Nusselt
Otto
Parsons
Planck
Prandtl
Rankine
Rumford
Stirling
Watt

Thermometry
Callendar
Celsius
Fahrenheit
Kelvin
Rankine

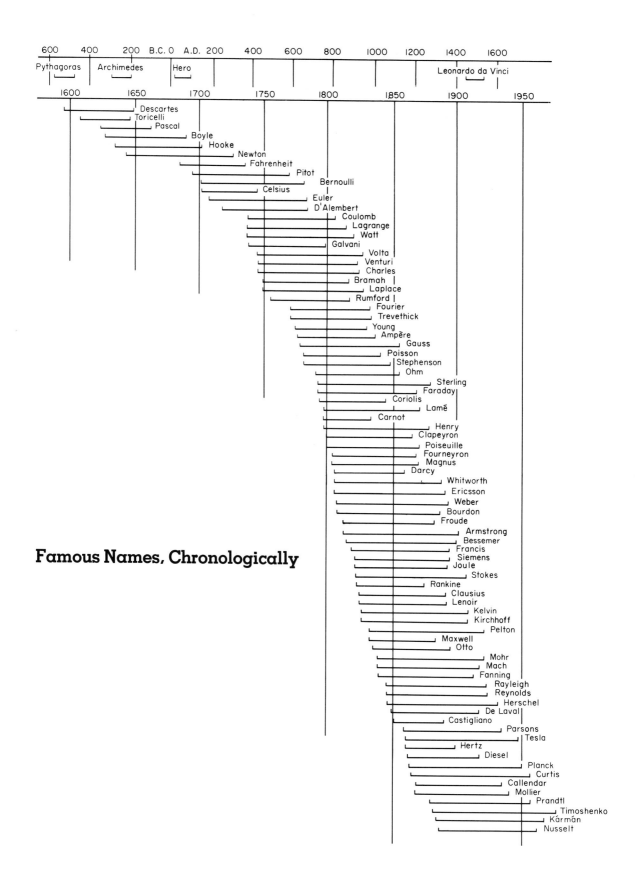

Famous Names, Chronologically

André Marie Ampère 1775-1836

French electrodynamicist, physicist

Ampère is associated, in the minds of both layman and engineer, with the science of electricity and particularly with the unit of electric current, which is named after him in both the cgs system and the 'Système International d'Unités (SI). Like many famous scientists, however, his interests were diverse and included mathematics, physics and chemistry. He studied philosophy and ethics under several famous philosophers of the day and showed remarkable ability in both subjects.

André Marie Ampère was born in the city of Lyons, France on 22 January, 1775. André Marie was a brilliant child, with a phenomenal memory, whose mathematical abilities developed early in life. Like another famous prodigy, Blaise Pascal, he was largely self-taught and when only thirteen years of age had produced a learned treatise on conic sections.

In 1793, when Ampère was eighteen, fate struck its first cruel blow. His father was declared an enemy of the Republic and was arrested by command of Collot d'Herbois and the notorious Fouché. He was tried, found guilty and guillotined in Lyons. In 1799, Ampère married and began to earn his living as a coach in mathematics, until he was appointed Professor of Physics at Bourg, about 40 miles from Lyons. After only five years of marriage, fate struck again with the death of his wife. This tragedy, following so shortly after the execution of his father, had a serious affect on Ampère's health, leading to a prolonged nervous breakdown from which he never really recovered. His illness was aggravated by the fact that he had disciplinary problems with his students.

Following the reading of one of his papers on mathematics, Ampère was appointed to the chair of mathematics at Lyons and later to a lectureship in mathematical analysis at the prestigious École Polytechnique in Paris. He took up residence in the capital and in 1808 became Inspector-General of the Imperial University. During the next twelve years, Ampère showed a remarkable understanding of all fields of science, including mathematics, chemistry and physics.

An important turning point came on 11 September 1820 when he witnessed a repetition of the classic experiments of Hans Christian Oersted on electromagnetism. Ampère was immediately attracted to this exciting field of science and he enthusiastically plunged into experiments which culminated in his demonstration of the induced force between two conductors when carrying an electric current. It was this phenomenon which was to lead to the invention of the dynamo by Michael Faraday in the 1830s. He formulated a precise definition of electric current and electromotive force and suggested that all magnetism may be due to electric currents. He came close to the relationship between the current flowing through a conductor and the applied electromotive force, stated later by Georg Simon Ohm (page 53).

While on a tour of duty as Inspector-General of the University of Paris, Ampère caught cold and contracted pneumonia. After a short illness, he died at Marseilles on 10 June 1836, aged 61.

Ampère was, by all accounts, a likeable and charming person in public but in private life he suffered from depression, deep religious turmoil and was tortured by doubts. He stands out as one of the giants in the science of electricity and Clerk Maxwell (page 48) rightly described him as 'the Newton of electricity.'

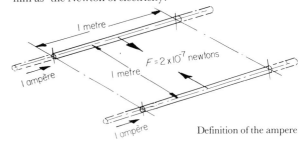

Definition of the ampere

Archimedes of Syracuse 287-212 B.C.

Greek mathematician and engineer

The name Archimedes usually brings to mind the amusing tale of his mad dash through the streets of Syracuse (in Sicily) shouting, "Heureka! Heureka!" meaning "I've found it! I've found it!". The incident is said to have occurred after he had stepped into an over-filled bath tub and suddenly realised how to measure the amount of gold in the King of Sicily's

new crown. The story may not be true, but it is a fact that Archimedes was the founder of the science of hydrostatics, which includes the concepts of specific gravity and the equilibrium of floating bodies. He established the principle, known as 'Archimedes Principle', that a floating body displaces a mass of water equal to its own mass.

Archimedes was also skilled in the science of geometry. He applied himself assiduously to the measurement of the lengths, areas and volumes of many geometrical curves, plane figures and bodies. He determined an accurate value for 'pi' and found expressions for the volumes and surface areas of the sphere, cylinder and cone. He also discovered the position of the centres of gravity of a number of geometric bodies.

Like many of his contemporaries, Archimedes assumed that the earth including its surface water was spherical and that there existed a force on all matter which acted towards the earth's centre! He was familiar with the principle of levers and often told his students "Give me a lever long enough and a place to stand and I will move the earth."

He invented the Archimedean screw pump, which consists of a helical screw inside an inclined tube, the end of which is submerged in water. Sufficiently rapid rotation of the screw produces a centrifugal pressure in the water which causes water to flow up the tube. This pump, often referred to as a 'Snail', was used extensively to irrigate the Nile Delta. When driven by treadmills through gearing, the pump was standard equipment in Roman mines.

When the Romans invaded Syracuse in B.C., Archimedes was called upon to organise the city's defences against a huge enemy fleet. When the attack took place, his monstrous catapults, hurling great stones and balls of Greek fire, caused havoc among the Roman troops. His cranes, mounted on the battlements overlooking the harbour, were used to drop boulders on the ships and some, fitted with grapnels, actually lifted the ships out of the water tipping the terrified occupants into the sea. The astonished Roman commander, Marcellus, called off the attack and retreated with the remnants of his fleet. He decided instead to lay siege to the city to bring about its fall.

When the Romans finally defeated the Syracusans in 212 B.C. and entered the city, Archimedes, instead of fleeing the city like the rest of the population, was busy with his calculations and stayed in his room. A Roman soldier burst into the room and, strictly against the orders of Marcellus, slew the great man.

Thus ended the life of a brilliant scientist and engineer who founded the science of hydrostatics and made many contributions to our understanding of the natural world.

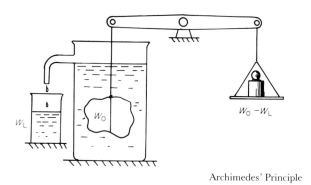

Archimedes' Principle

William George Armstrong, Baron 1810-1900

British industrialist and engineer

William George Armstrong was born on 26 November, 1810. His father William Armstrong, a corn merchant of Newcastle-upon-Tyne, formerly of Wreay near Carlisle, was the son of a shoemaker, who could not only read and write but had a good understanding of mathematics and took part to great effect in the local government of the City of Newcastle.

Young Armstrong suffered from ill health and was confined to his home for most of his early life. After attending several private schools in Newcastle, Armstrong was sent to Bishop Auckland Grammar School, County Durham. There he showed a great aptitude for mathematics and mechanics, and also experimented with hydraulic machines. In spite of his deep love for engineering, he had to leave school at the age of eighteen and, at the wish of his father he was articled to a solicitor friend of the family, Armorer Donkin, in Newcastle.

It was not until he was 37 that Armstrong abandoned law to indulge in his first love, engineering. With other Newcastle business men, he formed a company to produce hydraulic cranes which he had designed, at a small factory at Elswick on the north bank of the river Tyne, not far from Newcastle.

As he watched the few dozen men turn up for work on the first day, Armstrong could not have guessed that the factory would be employing 20 000 workers by the time of his death.

Initially, the Elswick factory produced hydraulic equipment for cranes, dock gates, bridges and hydraulic accumulators, but following the outbreak of the Crimean War, there was a great demand for improved field guns. Guns used at that time were little different from mediaeval cannon, so Armstrong designed and manufactured the famous 'Armstrong Gun' which had a rifled barrel, was breech-loading, and fired an elongated shell, thus making it the forerunner of all modern guns. He solved some difficult metallurgical problems and developed a method of making gun barrels by shrinking several wrought iron cylinders on to a hard steel liner.

Armstrong opened a shipyard at Elswick and in 1868 the warship, *H.M.S. Staunch,* was launched for the Royal Navy, fitted with Armstrong's guns. Over the next 15 years, 20 warships were built for the Royal Navy, and later, ships were built for navies all over the world, including China, the Argentine, Norway, Portugal, Turkey, the United States, Brazil, Rumania, Spain and Japan. Guns of all sizes and types were sold throughout the world for both land and naval use. In the Russo-Japanese war of 1902, ships on both sides fought each other with guns manufactured at the Elswick works.

The firm continued to prosper and played an important part in World War I supplying vast quantities of armaments of all types. A bitter controversy arose at one time between Armstrong's firm and that of Joseph Whitworth over the awarding of government contracts for armaments, but this was resolved when the two firms merged to form the new firm of Armstrong-Whitworth. In World War II, the factory at Elswick rallied to the national cause once more as Vickers-Armstrongs and produced great quantities of arms.

In 1887 Armstrong was created a Baron and retired to his beautiful estate at Cragside near Rothbury, Northumberland where he continued to work for up to 12 hours a day on electrical experiments until he was well into his 80s. His house was the first in the world to be lit by the electric lights invented by his great friend, Joseph Swan of Sunderland. It is interesting to note that the electricity was produced by a water-turbine-driven generator using water from a lake on his estate. This was the first hydro-electric installation in Britain.

Armstrong was one of the relatively small group of entrepreneurs who created industrial Britain, a man who managed to combine scientific brilliance with business sense. But it is difficult to decide whether he was a power for good or evil, as he shared with Krupp the reputation of being the leader in arms manufacture and also ranks in the world of engineering with men like Stephenson and Brunel. It is interesting to speculate how Britain would have fared in World War I if William George Armstrong's factory had not been there to supply 13 000 big guns, 100 tanks, 47 warships, 240 converted merchant ships, 1000 planes and three airships, together with 14 500 000 shells, 18 500 000 fuses and 21 000 000 cartridges.

Lord Armstrong died peacefully, aged 90, at his home on 27 December, 1900. His obituary in a Newcastle-upon-Tyne daily paper included the words, "There is something that appals the imagination in the cool application of a clear and temperate mind like Lord Armstrong's to the science of destruction."

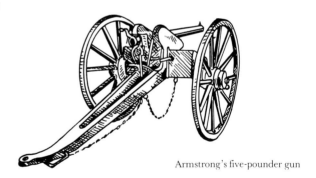

Armstrong's five-pounder gun

Daniel Bernoulli 1700-1782

Swiss mathematician and hydrodynamicist

Daniel Bernoulli was probably the most illustrious member of a wealthy merchant family which managed to produce no less than *nine* outstandingly brilliant mathematicians and scientists over three generations. His father Jean, was a great mathematician, famous for his 'Bernoulli Numbers' and his uncle Jacques, an important figure in the development of the calculus. His brother Nicolaus is noted particularly for his work on Probability Theory.

Born on 9 February, 1700 in Groningen, Holland, Daniel was the great-grandson of Jacques Bernoulli who, early in the 17th century had fled from his home town of Antwerp during a religious persecution against Protestants by the Duke of Alba. He first went to Frankfurt and finally settled in Basle in Switzerland where he became a successful merchant, taking up citizenship in 1622. His two gifted grandsons studied medicine and theology initially but later turned to mathematics. They travelled extensively in Europe, taking up various teaching appointments, and it was while Jean was Professor of Mathematics in Groningen that Daniel was born.

Like his uncle, Daniel studied medicine initially and was awarded a degree which was followed by the award of a doctorate for a thesis on the function of the lungs. Preferring mathematics to medicine, Daniel changed his career completely, studied for a degree in his newly chosen subject and in 1725 became Professor of Mathematics at St. Petersburg, a remarkable achievement by someone only 25 years of age. He worked there until 1733 but found the conditions at the Russian University so primitive that he was pleased to give up the post and return to the family home at Basle, where he went back to his first interest and accepted the post of Professor of Anatomy and Botany. A few years later Daniel was appointed Professor of Natural History, which position he retained until his retirement in 1777.

Daniel Bernoulli contributed greatly to the fields of mathematics and hydrodynamics; his work in mathematics included the solution of Riccati's differential equation and the development of partial differential equations. In conjunction with Euler and D'Alembert, he studied problems relating to the vibration of strings and also produced papers on the use of the trigonometrical series and on probability.

His *magnum opus* was undoubtedly his *Hydrodynamica* which he completed in 1733 after leaving St. Petersburg; however, the work was not published until 1738. Its 13 sections contain a complete history of hydrostatics and hydrodynamics and a kinetic theory of gases which is not unlike the more modern theory of Maxwell. It is in Part V of Section 12 that one finds the 'idea' of the conversion of energy of a flowing fluid from kinetic to pressure and vice-versa, but not an equation governing it. It was Bernoulli's close friend and co-worker, Leonhard Euler, who rigorously derived the so-called 'Bernoulli Equation'. *Hydrodynamica* also contains an interesting discussion on elastic fluids and is noted for its wealth of applications of the theory of fluid mechanics to practical problems. An interesting proposal, made years before its time, was that of screw propulsion for ships.

Bernoulli also made significant contributions to the study of elasticity of solids and suggested to Euler the use of calculus for the determination of formulae for the elastic curves of loaded beams. He was the first to derive equations for the lateral vibrations of prismatic bars.

One of the discomforting features of life in such an eminent family as the Bernoullis was that it was never possible to be sure which member of the family had actually made a discovery. There were several occasions when the brothers Jacques (Daniel's father) and Jean (his uncle) quarrelled over the question of priority of a new discovery. Jean even had a bitter disagreement with his son Daniel over the sharing of a prize from the Académie des Sciences in Paris.

Daniel Bernoulli died in Basle on 17 March 1782, aged 82, five years after his retirement from Basle University.

This great man made unquestionably a tremendous impact on the field of fluid mechanics and it is fitting that his fame should be perpetuated in the continual use of his name in all textbooks on that subject. It is of interest to note the family tree of the Bernoulli family shown here with the alternative German spellings of their forenames. The names are also numbered to avoid confusion.

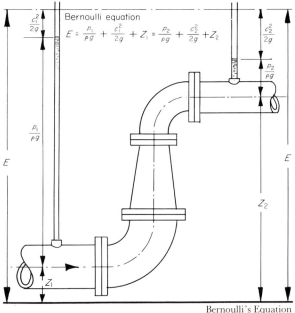

Bernoulli's Equation

$$E = \frac{p_1}{\rho g} + \frac{c_1^2}{2g} + Z_1 = \frac{p_2}{\rho g} + \frac{c_2^2}{2g} + Z_2$$

Jacques I (Jakob I)
|
Nicolaus I (Nikolas I)
|
Jacques II (Jakob II)   Jean I (Johann I)
| |
Nicolaus II (Nikolas II) Daniel Nicolaus III (Nikolas III)
                                          |
                                  Jean II (Johann II)
                                          |
                          Jacques III (Jakob III) Jean III (Johann III)

Henry Bessemer 1813-1898

British engineer

The inventive genius of Henry Bessemer led to the production of high quality steel on a vast scale in the middle of the 19th century, resulting in tremendous technological progress. Bessemer was fortunate enough to live to see the results of his work and receive the great honours, and profit, from his many inventions.

Henry Bessemer was born on 19 January, 1813 at Charlton, then a small village on the outskirts of London. His father, also a gifted engineer, had emigrated from London to Holland and then moved to France to work at the Mint in Paris. During the French Revolution he lost his job and his money and returned penniless to London. By dint of hard work and enterprise, qualities which his son was to inherit, Bessemer built up a business and bought a comfortable home with a splendid garden full of tulips, no doubt to remind him of his pleasant stay in Holland. The business included a foundry which produced high quality prints of artistic design made in exceptionally good quality alloys, and it was here that the young Henry gained valuable experience in the use of hand tools and machines, also with metals and their alloys.

Henry was educated locally even though his father, now quite well-off, wanted to send him to college for higher education. Henry pleaded successfully with his father to be allowed to work in the business. His father agreed and, at the age of 17, Henry had set up in business producing fusible alloys and art metals. He was a prolific inventor, his inventions including a machine for making Utrecht Velvet, a new method for making lead pencils upon which the modern production method is based, a type-casting machine, and a paper-embossing machine. There is a story concerning his sister who was decorating the cover of an album and wished to have gold lettering. Henry obligingly invented the first 'gold' paint for her, made from bronze powder.

There was hardly a field of technology which Bessemer did not explore—glassmaking, fuels, textiles, metals, and guns. It was, in fact, his interest in guns that led him to consider the possibility of improving the strength of iron to withstand the tremendous pressures involved. Having already carried out tests on rotating elongated shells for Napoleon III, he realised the limitations of cast iron for gun barrels, and began a series of experiments which were to lead to his greatest achievement, the production of steel.

His first experiments were carried out in 1855 using a fixed converter, then later with a tilting converter at Sheffield. An amusing story connected with the Sheffield tests concerns an experienced old iron-founder, who asked Bessemer, "Where be'ee goin' to put t'metal maister?". He was told, "Into that converter where I'll blow cold air through it to make it hot." The old man scratched his head and looked at him pityingly, "Ee, maister, if tha does it'll soon be all of a lump!".

After a number of initial problems, Bessemer succeeded in producing steel which was soon in demand all over the world for the new technology. The rapid growth of railways alone created a vast market and Bessemer made a fortune that enabled him to retire in 1869 when only 56, not that retirement meant his giving up inventing. A sufferer from sea-sickness, he designed a gyro-controlled circular saloon for a cross-Channel steamer for his personal use and, being a gifted draughtsman, he produced all the drawings down to the last detail. Unfortunately, the steamer was involved in two collisions and Bessemer lost £34,000 on the venture, a vast sum in those days.

Bessemer lived to the age of 85 in a beautiful house on Denmark Hill complete with farm, meadows and a large garden, where he died peacefully on 15 March, 1898.

One of the greatest inventors of all time, Henry Bessemer shunned specialisation and involved himself in many different aspects of technology. His development of steel was tremendously important even though the Siemens-Martin process was eventually to produce a greater quantity. As well as being a brilliant engineer, Bessemer also had a good head for business and made a considerable fortune out of his many interests.

Bessemer Converter

Eugéne Bourdon 1808-1884

French instrument maker

Most great men of science and technology are remembered for a theory or an invention producing a significant breakthrough in some field of engineering. In many cases, however, a name may be immortalised by a very simple invention. Such a common, yet important, instrument is the ordinary pressure gauge, and most engineers know that the heart of the pressure gauge is usually a 'Bourdon tube', named after an obscure Parisian instrument maker of the last century, Eugéne Bourdon.

For many years after James Watt had developed the atmospheric steam engine and Trevethick had introduced the high-pressure steam engine, a method was not available for measuring pressures of liquids, gases or vapours of much more than one atmosphere. James Watt had made a rather crude engine cylinder pressure indicator which showed how the pressure varied in the cylinder of his engines. This operated at low pressures in the region of one atmosphere. Another method of pressure measurement at that time was the manometer, but since one atmosphere pressure is equivalent to a 760 mm column of mercury in a U-tube manometer, it is obvious that such a device is useless for even moderately high pressures. The need for a compact and reliable instrument for measuring high pressures was great and it was left to the ingenious Eugéne Bourdon to produce in 1849, one of the most useful and versatile instruments available to the scientist and engineer.

Bourdon was born in Paris on 8 April, 1808, the son of a highly successful merchant. After an excellent education in Paris, Eugéne, who was expected to enter the family business, was sent to Nuremberg to learn the German language, a necessity in those days for a prospective French merchant. He returned to Paris after two years to work in the family business until his father's death in 1830. Then twenty-two, Eugéne immediately left the firm and went to work for an optician where he hoped to learn the art of instrument-making. After two years he started up his own instrument and machine shop in the back streets of Paris, then, in 1835, established a thriving business at 71 Faubourg du Temple.

Bourdon made a wide variety of scientific instruments and over 200 small steam engines, mostly for demonstration purposes. One beautiful example was presented to the Société d'Encouragement pour l'Industrie Nationale. His most important work however was the invention and perfection of the pressure gauge described in the patent of 18 June 1849, as a 'Manomètre Métallique sans Mercure pour l'Indiqués la Pression de la Vapeur dans la Chaudières'.–A Manometer without mercury to measure the pressure in Steam Boilers. Now known as the 'Bourdon Gauge', the instrument consists of a bronze or steel flattened tube bent into the shape of a 'C' with one end closed and the other end applied to the source of pressure. The closed end is connected to a mechanism which terminates in a pointer and a calibrated scale. When pressure is applied to the tube, it tends to straighten and this movement is magnified by means of gears to produce rotation of the pointer. Simply by altering the proportions of the tube a vast range of pressures can be accommodated, the full-scale pressure reading varying from fractions of an atmosphere to about 8000 atmospheres for both gases and liquids. Pressure difference can be measured by a differential gauge using two Bourdon tubes and a differential linkage. The movement of the tube may now be made to operate an electronic transducer.

Apparently Monsieur Bourdon produced his tube when he observed the distortion of a coil of lead pipe in a heater when the pressure was increased. As well as using the 'C' shaped tube, he made gauges with twisted flat tubes and helical coils of flat tube. There is still no satisfactory mathematical analysis of the phenomenon and design is largely based upon empirical evidence.

In 1851 Bourdon received the Legion d'Honneur after a highly successful demonstration of his gauges in the International Exhibition in London. Although his sons took over the business in 1872, Bourdon continued to work on the development of many types of instrument until his tragic death. On 29 September 1884 at the age of 76, Bourdon was perched precariously on his roof measuring wind speed with a new design of anemometer which incorporated a Venturi-tube when he lost his footing and fell to his death.

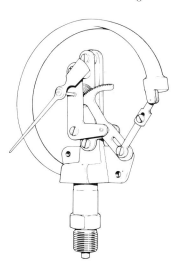

Bourdon Gauge mechanism (Budenberg Gauge Co. Ltd)

Robert Boyle 1627-1691

British natural philosopher

The name of Robert Boyle is familiar to all who have studied elementary physics in connection with the well known 'Boyle's Law' for gases. The law is usually stated: "For a given mass of a perfect gas at a given temperature, the volume is inversely proportional to the pressure," or in terms of a mathematical formula $pv = c$, where c is a constant. Such a precise statement would have appalled Robert Boyle who did not believe in quantitative results in science and was highly suspicious of the use of mathematics which he believed, with some justification, was used by many scientists to confuse others. The tremendous number of scientific investigations carried out by Boyle involved the absolute minimum of mathematics mainly because he wished it to appeal to a wide public.

Robert Boyle was born in Lismore, County Cork, Ireland on 25 January 1627, the seventh son of Sir Robert Boyle, Earl of Cork and Lord High Treasurer of Ireland. His financial situation was such that he did not find it necessary to work for a living but instead was able to follow his interest in science for the rest of his life. His family position also ensured his easy access to Court circles, so important to the up-and-coming young scientist, particularly when later he was involved in the founding of the Royal Society.

The young Boyle was educated privately, then at the age of eight entered Eton where he displayed a great flair for languages and was noted for his prodigious memory. After he had been at Eton for only four years, the Civil War broke out in England and Boyle was sent to the continent. He studied in France, Italy and Switzerland and became proficient at mathematics, modern history and geography as well as learning the gentlemanly arts of fencing and dancing. With the Irish Rebellion threatening his estates, Boyle returned first of all to London to settle his affairs and then settled in Stalbridge, Dorset, where he spent six years alternating between spells of scientific research, writing philosophy, and farming.

Following this period Boyle went to Ireland to repair his estates and then returned to England to live in Oxford where most of his experimental work was to be done. After 14 years at Oxford he moved to London where he remained until his death.

Robert Boyle's most important experiments were based upon the air pump experiments of Otto von Guericke, of Magdeburg Sphere fame. They included the boiling of warm water by reducing the pressure, and the weighing of air. The latter demonstration, carried out in front of the newly-formed Royal Society, caused great merriment among the non-scientific members including the King, Charles II, who lived up to his reputation as the 'Merry Monarch' by joining heartily in the laughter. Boyle's theory of heat was very similar to the modern kinetic theory in that it was based upon the motion of ultimate particles of a gas.

But perhaps Boyle's greatest contribution to science was his unusual approach to the study of chemistry. He attacked the Aristotelean view of chemistry and advanced the atomic theory in which all matter is assumed to be made up of primitive simple bodies which could be combined to make all types of matter found in nature. He pioneered the use of physical chemistry and insisted upon regarding a substance as an element until it could be resolved into simpler substances. It is difficult to understand how, in view of his changing of the whole nature of the subject, Boyle is barely mentioned in chemistry textbooks.

Dogged all his life by ill health, Boyle was deeply interested in diseases and wrote a whole series of books in which he advocated the scientific and experimental approach to medicine. He also studied thin films and carried out experiments on the nature of colours.

Towards the end of his life, Boyle was offered many honours and high positions which he steadfastly refused to accept. He adamantly declined the Presidency of the Royal Society in 1680. This great scientist and protagonist of the experimental method in science, died in London on 30 December 1691.

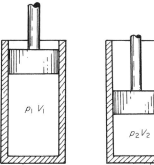

Boyle's Law

$p_1 V_1 = p_2 V_2$
If $T_1 = T_2$

Joseph Bramah 1749-1814

British engineer

In the North of England, one still occasionally hears the expression "It's a right Brammer," referring to something particularly well-made. It is highly unlikely, however, that the person using the expression will have any idea of its derivation, although it is an indication of the fame once achieved by one of Britain's greatest engineers, Joe Brammer, a Yorkshire ploughboy who later changed his name to Joseph Bramah in deference to London society. Bramah was the man who pioneered hydraulic power engineering with his famous press, made the 'burglar-proof' lock and invented the first effective flushing water closet.

The only known portrait of probably the most versatile engineer of the 18th century is to be found behind the President's chair at the headquarters of the Institution of Mechanical Engineers, London, with James Watt and George Stephenson looking on from each end of the room. It is a great pity that very few people know anything of Bramah and his work apart from having heard of the Bramah Press and Bramah lock.

Joe Brammer was born in the Yorkshire village of Stainborough near Barnsley on 13 April 1749 in a humble stone cottage which is still habitable. He was the son of a peasant farmer of extremely poor circumstances and seemed destined to follow the plough like his forebears when, at the age of 16 years, fate intervened and completely altered the course of his life.

During the Annual Feast held in the village when Joe was taking part in a jumping contest with the other village lads, he landed awkwardly and injured his ankle badly. During his convalescence, while indulging in his hobby of woodcarving, Joe made up his mind to forsake the plough and try to make a career out of his hobby. He served a seven year apprenticeship as a joiner and cabinetmaker and, finding the prospects poor at home, set out, gammy leg and all, to walk the 170 miles to London to seek fame and fortune. Thus began the career of the great engineer whose hydraulic press was to make possible the work of Robert Stephenson on bridges and the launching of Brunel's *Great Eastern*.

Bramah arrived in London at the age of 24 and, being a good carpenter, he had little difficulty in finding work, at the exceptionally good wage of 20 shillings a week. His skills took him into the houses of the wealthy and influential, where, in those days even the well-to-do used the most primitive privies, jocularly known as 'Jerichoes'. (The common people were still in the habit of discharging the contents of chamber pots from upper windows into the street below with a cry of "Gardy Loo!", a corruption of *"Gardez l'eau!"*, to warn passers-by.) During his spare time, Bramah designed and perfected the first really effective flushing water closet, vastly superior to the "Jerichoe". He saved half of his earnings for five years, for the amazingly high patent fee of £120; he must have had enormous faith in his invention.

When the patent was taken out, Brammer changed the spelling of his name to Bramah with the long vowel sounds then coming into fashion in London society in imitation of the Italian 'ah'. He was soon making considerable money out of his Water-closet upon a New Construction and by 1797 over 6000 had been sold. Bramah now had a factory where his chief assistant was Henry Maudslay who was destined to be one of the greatest inventors of machine tools. He decided that the hydraulic press should be developed at this factory.

The principle of the hydraulic press was known to the Ancient Greeks, and Gabriel Pascal had made a convincing looking sketch of one in 1664. Bramah's first attempts were doomed to failure because of the difficulty in sealing the ram against the extremely high pressure. It was only when Maudslay's genius produced the leather 'U' seal, which expands under the fluid pressure, that the press became a practical proposition.

Bramah's press turned out to be immensely useful for all manner of applications, baling cotton, flanging boiler plates and in bridge building and ship launching. It was particularly useful for the forging of large steel ingots where the steady pressure and precision control made it far superior to the steam hammer. It was also ideal for extrusion processes. Bramah invented a process for sheathing lead-covered cables using hydrostatic extrusion, a process reintroduced in recent years for cold extrusion of ferrous and other non-ferrous metals.

The wonderful partnership of Bramah and Maudslay also designed and manufactured the most accurate and ingenious of locks, far superior to any others made at that time. Bramah

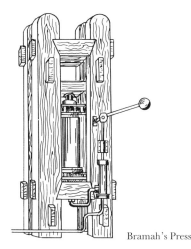

Bramah's Press

offered a prize of 200 guineas to anyone who could pick one of the locks which he described as 'burglar-proof'—it was 50 years before anyone opened it! Maudslay, who, incidentally, invented the screw-cutting lathe, made the high-precision tools required for the manufacture of the accurate parts for the locks and hydraulic presses. This was the first instance of mass production methods using precision machine tools.

Joseph was a prolific inventor, his other achievements including a beer engine, a machine for numbering banknotes, a planing machine, a stone saw and the first fountain pen. A patent of 1809 describes a remarkable hydro-pneumatic suspension system for carriages amazingly like some used on modern automobiles. The last patent he ever took out was for a special cement which prevented dry rot in timber.

Although the principles of hydraulic power had been known for 130 years Bramah laid the foundations for a whole field of technology. He was aware of the tremendous potential of hydraulics with its flexibility and its ability to transmit great power over a distance. It took another half century before men like Lord Armstrong began to exploit fully the great advantages of this new form of power.

The man who gave the world the hydraulic press died after catching pneumonia on 9 December 1814 and was buried in Paddington churchyard, London.

Hugh Longbourne Callendar 1863-1930

British physicist and engineer

About 180 years after Newcommen patented his first steam engine, Sir Charles Parsons produced his first steam turbine. Thirty years later, when the steam turbine was in wide use in warships and trans-Atlantic liners, Hugh Longbourne Callendar published his first 'Steam Tables'. It is difficult to imagine, when today one sees students of engineering referring so frequently to steam tables and charts, how great a boon they must have been to the designer of steam power plant in 1915. Callendar will always be remembered and honoured by the engineer for his out-standing contribution in the form of 'Callendar's Steam Tables', the culmination of many years of painstaking investigation into the properties of steam and water.

He is also famous for his great contribution to thermometry—the perfection of the platinum resistance thermometer which paved the way for precision measurement of the properties of gases and vapours—and for his many and varied researches in the field of steam and internal combustion engines.

Callendar was born at Hatherop, a village near Cirencester, in Gloucestershire, England, on 18 April 1863. He was educated at Marlborough School and then at Cambridge, where he studied both the classics and mathematics. He graduated initially in classics in 1884, then in mathematics in 1885 and became a Fellow in 1886. After serving as Professor of Physics at the Royal Holloway College, Egham from 1888 to 1893, Callendar went to Canada for five years as Professor of Physics at the McGill University, Montreal, where he met Howard Turner Barnes with whom he collaborated in experiments on calorimetry. In 1898, he returned to England as professor at University College, London and then at the Royal College of Science, later to be incorporated into the Imperial College of Science and Technology. He remained there as Professor of Physics until his death in 1930.

Callendar's main work was experimental in the field of heat and thermodynamics. He was particularly interested in the exact determination of the properties of water and steam, with a view to the better understanding of the processes in heat engines such as the reciprocating steam engine and the steam turbine. In 1900, he carried out a thorough study of the thermodynamics of steam and water, employing a modified form of the Joule-Kelvin equation, and then put forward his own equation for an imperfect gas which has proved to be very useful in determining the properties of steam. A paper describing this work was presented to the Royal Society in 1900 and formed the basis of his famous steam tables.

In addition to his sterling work on the properties of gases and vapours, Callendar demonstrated considerable ability as an engineer when he solved a number of practical problems associated with both steam and internal combustion engines. He carried out a great number of experiments on the flow of steam through nozzles and throttling devices, producing information of immense value to the steam turbine designer. He investigated leakage in piston engines and provided conclusive evidence of the phenomenon of supersaturation of steam during expansion.

In Montreal, Callendar, in collaboration with the talented Howard Turner Barnes, developed his method of continuous electric calorimetry in the apparatus so familiar to generations of school pupils who have studied elementary physics. This so-called 'Callendar and Barnes Apparatus' which eliminated the calculation of water equivalent of the apparatus and simplified radiation correction, was used to measure the specific heats of liquids and gave a much more precise value for the mechanical equivalent of heat.

The first work to be published by Callendar was in 1886 and dealt with his development of the platinum resistance thermometer as a new standard of temperature measurement. The improved design, which gave greater accuracy and stability, proved to be of great value in his work on steam properties, as it provided both the engineer and the research physicist with a thermometer not only of great precision, but also eminently suitable for recording purposes. A committee headed by Lord Rayleigh approved of the instrument, and the platinum resistance thermometer is now regarded as the international basis for a standard scale of temperature.

Callendar was involved in numerous other investigations. In conjunction with the staff of the Air Ministry, between 1925 and 1926, he published a paper on dopes and detonation in internal combustion engines and on the effect of adding anti-knock compounds to fuels. His talents covered a wide range, which was demonstrated when he invented a system of phonetic spelling and a form of shorthand which was actually in use in some British colonies.

Callendar received many honours, including the Watt Medal from the Institution of Civil Engineers, the Rumford Gold Medal from the Royal Society, and the Hawksley Gold Medal from the Institution of Mechanical Engineers, and was the first recipient of the Duddell Medal from the Physical Society. In 1920 he was awarded the C.B.E.

Hugh Longbourne Callendar died in London on 21 January 1930.

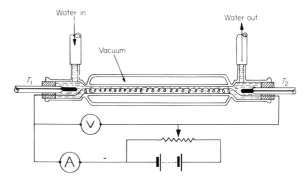

Callendar and Barnes apparatus

Nicolas Léonard Sadi Carnot 1796-1832

French physicist
and thermodynamicist

Sadi Carnot, as he is usually called, was born in Paris on 1 June 1796 into a distinguished French family which produced many famous men. His father Lazare, a brilliant General and Military Engineer who became Minister of War under Napoleon Bonaparte, later fell from favour after an abortive plot against the state and died in exile in Magdeburg. Sadi's brother, Lazare Hippolyte, was a famous Republican politician who at one time was France's Minister of Education. Lazare Hippolyte's son, Marie Francois, was also a politician who attained the exalted rank of President of the French Republic. He met a violent death at the hands of an Italian anarchist.

Sadi Carnot, the name Sadi being after the famous Persian poet whom his father greatly admired, entered the École Polytechnique in Paris, at that time regarded as the best scientific teaching establishment in the world, where he studied military engineering. His interest soon turned to the study of heat and in 1824 at the age of 28 he published his momentous work *Reflexions sur la Puissance Motrice de Feu* or Reflections on the Motive Power of Heat. It was a brilliant example of original thinking and led justly to his reputation as the 'Father of thermodynamics.' Carnot was thinking of the enormous industrial potential of the steam engine which, although it had been developed to a reasonable degree by British engineers, had not yet been subjected to a theoretical analysis. It was a tragedy for Sadi Carnot that his monumental work should lie practically unnoticed for 20 years until discovered by Lord Kelvin, who used the work, not for the development of heat engines, but as an aid to his calculations on the properties of substances.

In his *Reflexions,* Carnot showed that the most efficient thermodynamic cycle attainable is one in which all the heat is supplied at a fixed higher temperature, and all the heat rejected to the surroundings is at a fixed lower temperature. The cycle must therefore consist of two 'isothermal' processes with compression and expansion of the working substance taking place without the gain or loss of heat, i.e. adiabatically. It can be shown that the thermal efficiency of an engine operating upon such a cycle is the ratio of the difference of the higher and lower temperatures, to the higher temperature or $(T_1 - T_2)/T_1$.

The Carnot cycle thus forms the basis of comparison for all

practical engine cycles. The Rankine cycle, from which all modern steam plant has sprung, is a practical application of the Carnot cycle. The constant pressure cycle used for example in gas-turbines, and alternatively called the 'Joule, Stirling or Brayton Cycle', is based on the Carnot Cycles as are the Otto and Diesel Cycle's used for internal combustion engines.

It is remarkable that this brilliant work which laid the foundations for the later work of great scientists like Clausius and Kelvin, was produced before the declaration of the First and Second Laws of Thermodynamics. Although Carnot's great paper received little attention, apart from a passing reference to it by Clapeyron in 1834, he remains probably the most eminent thermodynamicist of his century. It was a great tragedy for science and technology when this brilliant man was lost to the world at the early age of 36. Towards the end of June 1832, Sadi Carnot fell ill, probably as the result of overwork and, after a partial recovery from an attack of cholera, died on 24 August.

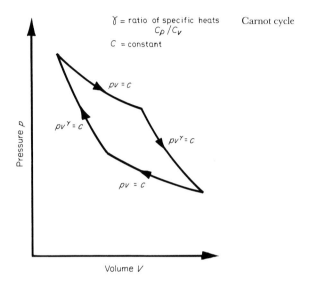

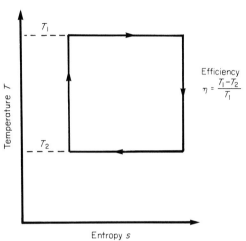

Carlo Alberto Castigliano 1847-1884

Italian structural engineer

Not a great deal seems to be known about Carlo Alberto Castigliano's short life. His name, however, is well-known to students of strengths of materials and structures in connection with 'Castigliano's Theorem' which relates to strain energy methods of analysis of structures. He was one of a group of outstanding Italian engineers, mathematicians and physicists of the 19th century responsible for the introduction of structural analysis based upon concepts of work and energy.

Castigliano's Theorem

$$\frac{\partial U}{\partial W_n} = x_n$$

$$\frac{\partial U}{\partial M_n} = i_n$$

U = Strain energy
W = Load
M = Moment
X = Deflection
i = Rotation

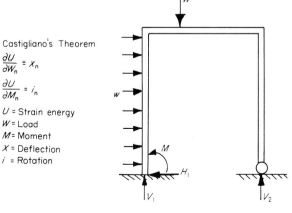

Castigliano's Theorem

Castigliano was born at Asti in Piedmont, Italy on 9 November 1847, and later studied at Turin. Graduating at the age of 19, he went to teach applied mechanics and engineering design at the Technical Institute at Terni in Umbria. After four years he returned to the Polytechnic in Turin and continued to work on the theories of elasticity which he had begun there as a student. He published his first paper 'Intorno si Sistemi Elastici,' at the age of 26. In it was included his First Theorem. This was followed about two years later by 'Intorno all' Equilibrio dei Sistemi Elastici,' including his Second Theorem.

The First Theorem makes it possible to determine the deflections in loaded beams and frames from a consideration of the strain energy stored in the members. He called this energy *lavoro molecolare*. The Second Theorem deals with the 'principle of least work'. Both theorems are also extremely useful for the calculation of thermal stresses.

Another contribution to engineering by Castigliano was an Engineers' Handbook which contained the theory of leaf and torsion springs and information on the design of masonry arches. It also included the design of a type of extensometer and the theory of water-hammer in pipes.

Castigliano was primarily an engineer and therefore greatly interested in the application of his theories to practical problems. He solved many complex problems in structures using his methods, which were vastly superior to the 'ad hoc' methods of previous investigators. His name, which trips so neatly off the tongue will be perpetuated wherever 'Structures' is taught.

Castigliano died at the early age of 37 on 25 October 1884 in Milan, having accomplished many great things in his short life.

Anders Celsius 1701-1744

Swedish astronomer

For some time there has been a certain controversy in Britain over the exact name of the scale of temperature which has been adopted to replace the Fahrenheit scale. What had always been referred to as the Centigrade scale is now to be called the Celsius scale in common with most of Europe. Fortunately both begin with the letter 'C' and we can safely use the symbol °C.

The Celsius scale of temperature was named after the Swedish astronomer who was born in Uppsala on 27 November 1701, where he received his primary education, then later attended the famous Uppsala University. In 1730 when he was 29 years old, Celsius was appointed Professor of Astronomy at Uppsala, a post he was to retain until his death.

He travelled extensively in Germany, France and Italy in connection with his interest in astronomy and presented a paper in Nuremberg on the first systematic observation of the Aurora Borealis which was made by him in the years 1716 and 1732. He published a number of other works on astronomy, including one on the relative brightness of stars, but none of these was of great importance. In 1740 he travelled to Lapland with several famous scientists to measure the meridian arc, while another group went to South America to determine the length of a degree of latitude near the Equator.

The most memorable incident in the life of Anders Celsius, was when he presented a memoir, read before the Swedish Academy of Science in 1742, in which he set out to persuade scientists throughout Europe to adopt one temperature scale and to convince them that the most suitable was the Centigrade or Centesimal scale. Most of Europe adopted what came to be called the Celsius scale with the exception of Britain, and other English-speaking countries, who used the temperature scale named in honour of the inventor of the mercury thermometer, Daniel Fahrenheit.

It is not known generally that the original Celsius scale had as its zero the *boiling* point of water and a value of 100° at the freezing point. The scale was inverted in 1750 by a physicist called Strömer, with goodness knows how much confusion, so that 0°C corresponded to the melting point of ice at atmospheric pressure and 100°C to the boiling point of water at atmospheric pressure. The degree Celsius is not strictly the S.I. (Systéme International) unit of temperature measurement but the 'kelvin' (K) named after Lord Kelvin.

Celsius temperature scale:
  0°C = Melting point of ice at a pressure of one atmosphere
  100°C = Boiling point of water at a pressure of one atmosphere

Absolute temperature in kelvins:
  $K = °C + 273.15$

Conversion to Fahrenheit:
  $°F = (°C \times 1.8) + 32$

The kelvin is the unit of the 'thermodynamic' or 'absolute' scale of temperature and the relationship between them is K = °C + 273.15, so that 0°C = 273.15 K, and 100°C = 373.15 K. Thus a temperature change of 1°C is equal to 1 K or 'one kelvin'.

Anders Celsius died at Uppsala on 25 April 1744 at the age of 43. His everlasting fame rests on his suggestion that scientists should adopt the same temperature scale. With help from Strömer, who turned the scale upside down, and the passing of two and a half centuries, we may yet see its universal adoption.

Jacques Alexandre César Charles 1746-1823

French physicist

Early in their studies of physics at school, pupils are introduced to two famous laws connected with what is referred to as a 'perfect gas'. One is named after an Irishman the other after a Frenchman. The first is of course Boyle's Law and the second is Charles' Law, the latter's name usually being anglicised in its pronunciation. Although the Laws are taught together and combined into what is called the Characteristic Gas Equation, Charles' Law originated 140 years after Boyle's. Another interesting fact is that neither man actually wrote down the Laws.

Jacques Alexandre César Charles was born into a wealthy family in the Loiret district of France at Beaugency on 12 November 1746 and being entirely free of any financial burdens, was able to devote his time entirely to the study of the sciences. As a young man he was fascinated by experimental science and in particular was interested in the wonderful experiments of the American scientist Benjamin Franklin who flew his kite in a thunderstorm to collect electricity. Charles was able to build up an impressive collection of scientific apparatus and became celebrated throughout the whole of France, among both scientists and the general public alike, for his brilliant lectures illustrated by numerous ingenious demonstrations.

After a deep study of the work of Priestley and Cavendish on the preparation of gases, Charles began a series of experiments on gases and filled balloons with hydrogen, a gas which had been isolated not many years before by Cavendish. On 1 December 1783, after a number of unmanned ascents, he ascended from the Tuileries Gardens, Paris in a hydrogen-filled balloon and reached the amazing height of 3 kilometres, travelling a distance of over 44 kilometres. This historic flight took place only ten days after the first manned ascent of a Montgolfier hot-air balloon.

Charles later experimented with balloons of elongated shape which he unsuccessfully tried to propel. Charles became the darling of the French populace as a result of his intrepid flights and was given a pension by the King, the ill-fated Louis XVI, who invited him to the Louvre Palace where he was allowed to establish his apparatus.

Nine years later on 10 August 1792, when the bloodthirsty Revolutionary mob invaded the Tuileries, Charles was one of the many aristocrats to be confronted by them. He saved himself an untimely death on the guillotine by having the presence of mind to remind the howling mob of his exploits in his 'Charliéres', as his hydrogen-filled balloons were known. These had so captured the imagination of the French public, that his life was spared.

In 1802 Charles, safe in his laboratory once more, carried out experiments on the dilation of gases. These experiments were similar to those of his contemporary Gay-Lussac but, unlike Gay-Lussac, he did not publish his results. Thus it was that his law was first mentioned in a publication by Gay-Lussac which stated "If the volume of a given mass of gas is taken as unity at zero temperature (0°C), then, at constant pressure the volume will increase by 1/267 for each degree rise in temperature". In other words, the volume of a gas is proportional to the absolute temperature if the pressure is contant. It is interesting to note that Boyle's Law relating pressure and volume was formulated as far back as 1662. The Characteristic Gas Equation $pV = mRT$, combining the two laws had to wait another 40 years to be formulated by Clapeyron. Both Laws only apply to what is known as a 'perfect gas'. They are of course empirical and accurate over only a limited range. Both Gay-Lussac and Charles had unwittingly discovered the existence of an absolute zero temperature 50 years before the accredited discovery of Lord Kelvin. Their results implied an absolute zero of —267°C instead of about —273°C.

Charles deserves fame not only for his extremely valuable contribution to the study of the properties of gases, but also to

---

Charles' Law: $V = V_0 \left(\dfrac{273+T}{273}\right)$

V = Volume of a given mass of gas
V_0 = Volume of the gas at 0°C
T = Temperature of gas (°C)

Bénoit Pierre Émile Clapeyron 1799-1864

French civil engineer

The name Clapeyron occurs in two completely different branches of engineering. It can be found in textbooks on Strengths of Materials in connection with strain energy methods to determine deflections of continuous beams, also in the field of thermodynamics where it is coupled with that of Clausius in the study of the evaporation, sublimation and melting of substances.

Bénoit Pierre Émile Clapeyron was born in Paris on 26 February 1799. He graduated initially from the École Polytechnique, then from the École des Mines in Paris in 1820 with his great friend Gabriel Lamé, well-known for his formulae for stresses in thick cylinders under pressure. As young engineers of great promise, they were recommended to the Russian Government by the French authorities to assist in the Russian bid to enter the new world of science and technology. They went to St Petersburg (now Leningrad) to teach applied mathematics and physics at the recently founded Institute of Engineers of Ways and Communications. In addition to teaching they soon found themselves engaged on the design of roads, public buildings, tunnels and the first suspension bridges to be built in Europe. One investigation undertaken by them was on the stability of the dome and arches of the Cathedral of St Isaac then being built in St Petersburg. This formidable range of work was undertaken by men still in their early twenties!

As a result of the July Revolution in 1830, relations between France and Russia deteriorated to such an extent that Clapeyron and Lamé, were forced to return to France. They both became involved in the construction of railways in the vicinity of Paris, one of which, the St Germain Railway, required a special design of locomotive to contend with the exceptionally steep gradients. Clapeyron travelled to England to the Newcastle-upon-Tyne locomotive works of Robert Stephenson to negotiate the purchase of several special locomotives from this famous man. Stephenson declined the order because of a number of design difficulties involved. But a versatile engineer like Clapeyron was not easily deterred–he designed the locomotives himself and had them made at the firm of Sharp and Roberts.

Although deeply involved in the solution of practical engineering problems, Clapeyron continued to teach at the École des Ponts et Chaussées where he ran a special course on steam engines. By all accounts he was an excellent teacher who managed to combine great theoretical knowledge with practical application. At about this time he wrote a useful paper on the regulation of slide valves in steam engines and, in contrast to such a down-to-earth subject, did work leading to the well-known 'Clapeyron-Clausius Equation'. This was an explanation of Carnot's principle as put forward in his much-neglected *Réflexions*.

On the subject of Elasticity, Clapeyron followed the work of Navier on Continuous Beams and introduced his own famous 'Equation of Three Moments' so familiar to engineering students taking advanced Strengths of Materials. This equation, useful for the solution of beams resting on more than two supports is particularly important to the structural engineer. One of its first practical applications was in the investigation of the strength of Robert Stephenson's Britannia Bridge over the Menai Straits in Wales. Clapeyron's Theorem deals with Strain Energy methods and states that 'the sum of the products of the forces and their displacements acting on a body, is equal to twice the strain energy stored in the body.'

Clapeyron was an engineer of outstanding ability who also contributed enormously to theoretical aspects of technology. He was elected a member of the French Academy of Science in 1858 and continued work at the École des Ponts et Chaussées and at the Academy, until his death on 28 July 1864.

$$M_A L_1 + 2M_B(L_1 + L_2) + M_C L_2 - (w_1 L_1^3 - w_2 L_2^3)/4 = 0$$

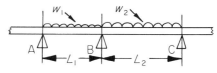

Clapeyron's equation of three moments

Rudolf Julius Emmanuel Clausius 1822–1888

German theoretical physicist

The engineering student's association with Rudolf Clausius results from his introduction to the world in 1865 of that most mysterious of fluid properties, entropy, to the everlasting bewilderment of students. The symbol he used for entropy was S, with heat flow represented by Q and temperature by T. At about the same time, the Scotsman Rankine, introduced the property as 'thermodynamic function', for want of a better term, with the symbol φ which was used until quite recently in Britain, until it was replaced by the symbol s.

Clausius was born at Coslin, Pomerania, on 2 January 1822. He studied at the University of Berlin and obtained his degree at Halle in 1848. He became Professor of Physics in 1850 at the Royal Artillery and Engineering School in Berlin and was later appointed Professor at Zurich, Wurzburg and finally at Bonn.

In the mid-nineteenth century, Lord Kelvin had commented upon the dilemma that, although Carnot's Theory was acceptable as far as its consequences were concerned, his use of the calorific theory of heat was inconsistent with the mechanical theory of heat. Clausius showed, in his epoch-making paper entitled *'Über die Bewegende Kraft der Wärme'*, that the problem could be resolved if it was assumed that heat could not by itself pass from a colder to a hotter body. In the following year, 1851, Lord Kelvin formulated the Second Law of Thermodynamics.

It was not until 1865 that Clausius first used the word 'entropy' which he derived from the Greek 'en' meaning 'in', and 'tropos' meaning 'transformation'. It is used to express the dissipation of energy in terms of 'increase in entropy'. In any practical process the entropy must increase, in fact the entropy of the whole universe tends towards a maximum. As was to be expected there was a bitter controversy by the most eminent scientists of the day over such a revolutionary concept but it was finally accepted.

The practical use of entropy came with the introduction of the Temperature–Entropy or T-s diagram introduced by Belpaise in 1873 and used by MacFarlane Gray in 1889. With the advent of the steam turbine, Mollier introduced the Enthalpy-Entropy or h-s diagram. Both of these gave a new insight into the working of a heat engine at a time when they were developing at a tremendous rate. When engine cycles were represented on these diagrams, a more complete picture of the cycle of operation could be obtained by the engineer than was possible with the p-v diagram.

Clausius died at Bonn on 24 August, 1888. He was undoubtedly one of the most original theoretical physicists of the 19th century, unsurpassed for his profound intuition and the ability to make far-reaching deductions without the necessity for complicated mathematics. His concept of 'entropy' gave a tremendous impulse to the development of the heat engine.

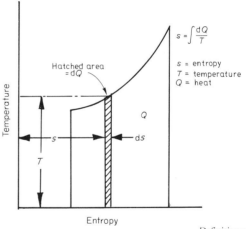

Definition of entropy

Gaspard Gustave de Coriolis 1792–1843

French mathematician and physicist

At some time or another the engineering student will be introduced to 'Coriolis' Component' in his Applied Mechanics lectures. It is the acceleration which acts sideways on a rotating body when it moves to a larger radius, or which produces the force which knocks a person sideways when he tries to walk towards the edge of a roundabout.

Coriolis was a French mathematician and physicist who devoted his energies to the applications of the laws of mechanics and who gave us the terms 'work, or in French *'travail'*, and 'kinetic energy' or *'force vive.'*

Gaspard Gustave de Coriolis was descended from an old Provençal family ennobled in the 17th century. He was born in Paris on 21 March 1792 into troubled times, the son of Jean Baptiste Elzéar and Marie Sophie (de Maillet). His father was a loyalist officer serving under Louis XVI and, to escape the Reign of Terror, had to flee to Nancy where later he became an industrialist. In 1808 the young Coriolis entered the Napoleonic École Polytechnique, a college for top civil servants. After leaving college he spent several years in the army as an Engineering Officer serving first of all in Meurthe-et-Moselle and then in the Vosges Mountains with a mountaineer corps. In 1816 Coriolis' father died leaving the hitherto considerable family fortune in such a poor state that Gaspard had to take up employment as a tutor at the École Polytechnique. He was recommended for the post by the celebrated mathematician Augustin Cauchy who inspired the young man to devote his life to the study of science.

In 1829 Coriolis accepted the Chair in Applied Mechanics at the newly-founded Centrale des Art et Manufactures. He also assisted the great Navier in the teaching of applied mechanics at the École des Ponts et Chaussées and succeeded him as Professor of Mechanics in 1836. Two years later he became Director of Studies at the École Polytechnique, a post in which he excelled.

Coriolis was intensely interested in the education of scientists and engineers and was particularly keen on the application of mathematics and science to engineering problems; he was also very popular with the studies both as a lecturer and as Director. He is remembered to this day by students who refer to the water coolers he had installed in the classrooms for their benefit, as 'Corios.' They may still be seen in the École Polytechnique.

Bad health had dogged Coriolis all his life which is probably the reason for his not being married. After his appointment as Director at the École Polytechnique his physical condition grew steadily worse and he died in Paris on 19 September 1843 at the age of 51 and was buried in Montparnasse cemetery.

Coriolis was a great believer in the application of classical mechanics to actual machines and gave the terms 'work' and 'kinetic energy' their modern meanings. His work included papers on the construction of roads, the theory of vortices such as whirlpools, compound centrifugal forces, and a light-hearted paper on the mechanics of billiards titled *Théorie Mathématique des Effets du Jeu de Billard.*' His name will of course be linked mainly with his discovery of the 'Coriolis Component of Acceleration' previously overlooked. This acts on an object when it moves along a rotating path and is at right angles to that path. A typical example is that of a pin sliding in a rotating slotted link. The so-called 'Coriolis Effect' on the motion of the Earth's oceans can be explained in terms of relative velocities. In 1963, more than a century after his death, a French oceanographic research ship was named *'Coriolis'* in honour of this great French scientist.

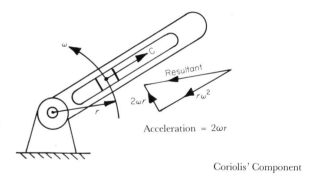

Coriolis' Component

Charles Augustin de Coulomb 1736-1806

French military engineer and physicist

The name Coulomb is asociated by the engineering student almost exclusively with the unit of electric charge, or quantity of electricity, and with his two famous laws concerning electrostatic and magnetic attraction; the terms 'Coulomb Friction' and 'Coulomb Balance' are not so well known. Charles Augustin de Coulomb however played an extremely important role in developing the hitherto vague and fragmentary knowledge of electricity and magnetism into a precise science by introducing great laws and by his brilliant experimental work. He developed electrostatics into a complete science and gave magnetism a firm foundation upon which men like Gauss and Weber could build. But Coulomb's genius as a military engineer is not generally known.

A century after Isaac Newton had published his Law of Gravitation, Coulomb discovered his completely analogous laws for electricity and magnetism. He showed that the force of attraction between electric charges and magnetic poles was

proportional to the products of the charges or pole strengths and inversely proportional to the square of the distance between them. These apparently simple laws required a tremendous amount of experimental skill and extremely accurate measurement of the minute and fleeting quantities involved.

Charles Augustin de Coulomb was born on 14 June 1736 in the French town of Angoulême, about 70 miles from Bordeaux. His father Henry, who had been a soldier, held a rather insignificant government post, his mother was related to a very wealthy family called de Sénac. When Charles was a boy his father was transferred to Paris and his mother was determined that he would become a doctor of medicine. She managed to get him into an exclusive private school intended for the sons of the nobility, called the 'College des Quatre Nations.' Apparently he attended only as a part-time student as he was referred to as a 'martinet' (A house martin flitting from place to place!).

Much to his mother's annoyance, his father did not seem to be very interested in Charles' education, but young Coulomb began to go to lectures by the famous mathematician Le Monnier at the 'College Royal de France' and was so impressed that he decided to become a mathematician. His plans however were thwarted when Monsieur Coulomb was made penniless by a series of bad financial speculations. Charles, more-or-less disowned by his mother, went to live with his impoverished father at Montpellier, where more affluent relations gave them assistance including an introduction for Charles to the thriving scientific circle there. It is highly probable that he attended the Jesuit College in Montpellier. At the age of 21 he presented several memoirs on Astronomy and Mathematics.

Coulomb was now faced with choosing a suitable and profitable career. The possibilities available for a young bourgeois at that time were the church, the army or the civil service. He joined the army and hoped that in the engineering corps he would be able to practise his mathematics. He entered the College of Mezieres where he received a salary of 600 livres and a resplendent uniform. The engineering training was excellent and included calculus, geometry, dynamics and hydrodynamics, with a great deal of practical training in the building of bridges, roads and fortifications..

In 1759 Coulomb graduated and two years later, young and inexperienced, was sent to Martinique in the West Indies to build fortifications in the harbour, to replace a brother officer who was ill. He was to be entirely responsible for the building of Fort Bourbon using a workforce of 1200. The conditions were appalling and many men were ill and quite a number died. Coulomb himself was seriously ill several times and never fully recovered. This experience helped to develop the engineer in Coulomb, a man who is generally thought of as a physicist. He wrote "...I was often discovering how much all the theories founded upon hypothesis, or the experiments carried out in miniature in the laboratory, were insufficient guides in practice."

After about eight years in Martinique Coulomb returned to France and, continuing in the army, wrote memoirs on many engineering subjects including the design of dry docks, hydraulics and a proposal for a diving bell to facilitate underwater construction work. During this important period in his life, Coulomb wrote his great work on statics and dynamics, *The Theory of Simple Machines.* He did some useful work on the torsion and elasticity of wires, presumably when working on his torsion balance. He also carried out a thorough investigation on friction, and was the first to realise that the friction force is proportional to the normal force between contacting surfaces and practically independent of the contact area. In 1779 Coulomb returned to Paris to settle the affairs of his mother who had just died and in 1781, he was honoured by being elected to the Académie des Sciences.

It was during the five years leading up to the Revolution of 1789 that Coulomb wrote his famous memoirs on electricity and magnetism which included his work on the torsion balance. He also studied the effect of the earth's magnetic field on magnets and established the equations of motion for a magnet in a magnetic field using his torsional oscillation method.

With the Revolution, Coulomb resigned from all his posts including that of Lieutenant Colonel in the army and was one of the 'nobles' who were expelled from Paris. He retired to a small estate in Blois with his friend J.C. Borda, the hydraulics engineer after whom the 'Borda re-entrant orifice' is named, and settled down to a quiet life of scientific investigation. He returned to Paris in 1795 on the invitation of Napoleon and was appointed one of the Inspectors-General of Public Instruction in 1802.

Coulomb continued to devote his life to scientific work until his death in Paris at the age of 70 on 23 August 1806.

Coulomb's laws for electricity and magnetism

$$F \propto \frac{Q_1 Q_2}{\epsilon d^2} \; , \; F \propto \frac{M_1 M_2}{\mu d^2}$$

where: Q = Electric charge
ϵ = Relative permittivity
M = Magnetic pole strength
μ = Relative permeability
d = Distance between poles or charges
F = Force of attraction

He was certainly one of the greatest physicists of the 18th century as well as being an accomplished engineer. His name is immortalised as the name of the S.I. unit of electric charge or quantity, defined as the quantity of electricity passing through a conductor in one second when a current of one ampere is flowing.

Charles Gordon Curtis 1860-1953

American engineer

The American engineer, Charles Gordon Curtis, is best known for his invention of the velocity-compounded impulse steam turbine, usually referred to as the Curtis Turbine in 1896. This type of turbine and the reaction turbine, invented by Charles Parsons, have continued to compete with one another in the field of power generation. Both types of turbine, however, have been used in combination in the so-called 'impulse-reaction' turbine where the initial pressure drop takes place in a two or three-row 'Curtis Wheel' and the remainder in a number of reaction stages of the Parsons type.

Curtis was born in Boston, Massachusetts on 20 April 1860, the son of George Ticknor and Louise Curtis, and began his technical education in 1877 at Columbia University where he qualified in 1881 as a civil engineer. He then decided to study law and after obtaining his LL.B. at the New York Law School in 1883, practised as a patent lawyer for eight years, an experience which was to stand him in good stead in later years.

His first business venture was to found the C and C Electric Motor Company which made electric motors and fans. He played a major part in the development of the Curtis Electric Manufacturing Company, and it was as President in that firm that invented, developed and marketed the Curtis Turbine, the rights of which he later sold to the General Electric Company, who introduced the turbine to the navies of the United States of America and other countries.

Curtis received many honours including the Count Rumford Gold Medal from the American Society of Arts and Sciences. He died on 10 March 1953 and was buried at Westchester, New York.

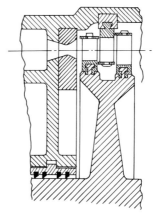

Curtis turbine

Jean le Rond D'Alembert 1717-1783

French mathematician and philosopher

Famous for his 'Principle' so well-known to students of applied mechanics, Jean le Rond D'Alembert began life in a humble and inauspicious manner. The illegitimate son of a noblewoman and an officer of the French Army, he was abandoned near the church of St Jean le Rond in Paris in November, 1717 a short time after his birth. Found and raised by the wife of a Paris glazier, who named him after the church, he received anonymous support of his natural father in the form of an annuity which he was to receive throughout his education.

Jean le Rond D'Alembert was educated at the Jansenist College of Mazarin where he studied law and then medicine, both of which he gave up in favour of mathematics in which he excelled. After concluding his education he returned to his beloved foster-parents to live with them for thirty years.

D'Alembert presented a number of papers to the

Académie de Sciences mainly in the field of calculus, and was made a member at the age of twenty-four. In 1743 he published his *'Traité de dynamique'* in which the well-known 'D'Alembert's Principle' appears as a method for reducing dynamics problems to ones of statics, by the introduction of forces due to the accceleration of the masses involved.

It is not generally known that D'Alembert next applied his Principle to the flow of fluids as described in his *'Traité de l'équilibre et du mouvement des fluides'*. He also carried out intensive investigations in the fields of vibration and sound, applied partial differential calculus to the analysis of wind flow, and solved the problem of the precession of the equinoxes. Although he received lucrative offers from the rulers of Prussia and Russia to work in those countries, D'Alembert refused to leave his native Paris.

In the middle of the 18th century D'Alembert and other famous mathematicians and scientists, including his close friends Leonhard Euler and Daniel Bernoulli, were deeply interested in the phenomenon of fluid resistance. The Berlin Academy sponsored a prize competition for the best paper on the subject and D'Alembert submitted his work which he later withdrew when the competitors were asked for experimental evidence to support their solutions. Instead he used the paper as the basis for his *'Essai d'une nouvelle théorie sur la resistance des fluides'* published in 1752, which contained his 'paradox'. This concerns the flow of a fluid past an oval shape for which he calculated the pressure forces exerted on the surface of the body, only to obtain the paradoxical result that the net force was zero, contrary to actual experience. D'Alembert wrote, "This, I do not see, I admit, how one can explain satisfactorily by theory the resistance of fluids–the theory in all rigour, gives in many cases zero resistance, a singular paradox which I leave to future Geometers for elucidation."

It was at that time that D'Alembert was involved with the renowned writer Denis Diderot in the preparation of the latter's mammoth *Encyclopédie ou Dictionnaire Raissoné des Sciences, des Arts et des Métiers*, comprising 17 volumes, with an additional 18 volumes of tables, illustrations and supplements. D'Alembert contributed many articles on literature, mathematics, philosophy and music. In 1755 he was elected to the Académie Française and became its permanent secretary in 1772.

Jean le Rond D'Alembert died, probably in Paris, in October 1783, to be for ever honoured as a brilliant mathematician and for his great contribution to mechanics and hydrodynamics.

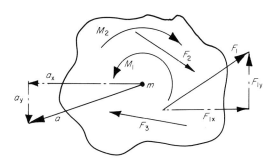

$\Sigma F_x = \Sigma(dm) a_x = ma_x$
$\Sigma F_y = \Sigma(dm) a_y = ma_y$
$\Sigma M = 0$

D'Alembert's principle

Henri Philibert Gaspard Darcy 1805-1858

French civil engineer

The name Darcy is best known to engineers, in particular civil engineers, in connection with 'Darcy's Formula' and 'Darcy's Law'. The first of these deals with the flow of fluids through pipes and channels, and the second with flow of fluids through porous media.

Henri Philibert Gaspard was born in Dijon, France, on 10 June 1803 and, after a primary education in Dijon, studied in Paris, specialising in the mechanics of fluids. He returned to his home town and took up employment with the local water authority in 1825 to be responsible for the design and construction of a new water supply system. It was in connection with this scheme that he carried out many investigations into the pressure losses in pipes, channels and filter beds. He was first to suggest that, for turbulent flow in pipes and channels, the loss in pressure, or head, is partly dependent upon the degree of roughness of the walls of the pipe or channel, a factor that was rather surprisingly neglected by previous investigators. Darcy's Formula for

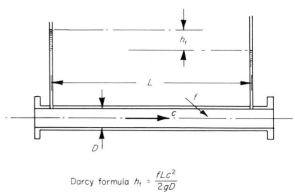

Darcy formula $h_f = \dfrac{fLc^2}{2gD}$

head loss includes a friction factor which is a function of the roughness and another factor later to be known as Reynold's Number.

Darcy also carried out some useful work on filtration which is described in his book of 1856 on the fountains of Dijon. He patented a filtration system which had many novel features, and greatly improved Henri Pitot's famous velocity measuring tube by introducing a very small diameter mouth which reduced interference and oscillation of the liquid column.

In the study of ground-water hydrology, Darcy showed that for a flow of liquid through a porous medium, for example, water flowing through sand, the flow is directly proportional to the pressure difference. It is therefore similar to the flow calculated by using Poiseuille's equation for pipes in which there is laminar flow. The theory, known to civil engineers as Darcy's Law, also applies to large pipes full of sand such as occur in filtration systems. Darcy collaborated with another Frenchman, Henri Émile Bazin who is known for his work called *Recherchés Hydraulique* which includes what is called 'Bazin's Formula'.

Darcy moved to Paris after a notable career in his birthplace, Dijon, where he died on 3 January 1858. It is fitting that the illustrious name of this pioneer of fluid flow in pipes and conduits should be remembered in the pipe-friction loss formula.

Carl Gustav Patrik de Laval 1845-1913

Swedish inventor and engineer

Like Thomas Alva Edison, de Laval, the Swedish engineer possessed an inventive talent which was unlimited in both energy and variety. Out of his many inventions his greatest achievement was undoubtedly the steam turbine named after him, which was first marketed in 1890, seven years before Sir Charles Parsons' *'Turbinia'* created a sensation at the Jubilee of Queen Victoria. His name is also connected with the de Laval separator used for cream making and the recovery of lubricating oil.

De Laval was born in the small town of Orsa in the Kopparberg mining district in Sweden on 9 May 1845. His family had emigrated from France in the 17th century during the persecution of the Huguenots. He was educated initially at the Technical Institute in Stockholm and then at Uppsala University, where he gained his Batchelor's degree in 1868 and his Doctor's degree in 1872. He worked for a time as an engineer in the steelworks of Klösters Bruck but soon started up his own engineering firm where he was eventually to employ two hundred engineers working on his numerous inventions.

In 1878 de Laval invented his cream separator which worked on the centrifuge principle and was driven by a very high speed turbine of his own design. The separator was a great success and revolutionised butter making. It was sold worldwide, and is still being marketed by the Laval Separator Company. This machine was improved in 1890 by the inclusion of von Bechtolsheim's 'Alfa' discs which gave more rapid and complete separation. One version of the machine is used to recover lubricating oil by centrifuging to remove dirt and water. He invented and marketed many other devices for the dairy industry, including a vacuum milking machine which was not perfected until the year of his death, 1913.

De Laval's greatest achievement, his steam turbine, was an attempt to produce a large quantity of power from a single row of moving turbine blades and thus achieve an extremely compact machine. He managed to solve the many design problems in spite of the lack of reliable data then available on the properties of steam. The turbine had convergent-divergent nozzles to produce a supersonic velocity of over 1000 metres per second and rotor speeds of up to 20 000 revolutions per minute. To achieve such high speeds, special attention was given to the design of the rotor which had a hyperbolic profile to give constant strength and was mounted on a very flexible shaft running well above the critical speed of whirling.

Attempts were made by de Laval to increase the power of his turbine so that it could be used for marine applications. In this attempt he was unsuccessful and eventually Parsons won the battle. He even built an experimental 15 hp turbine to propel a small launch but it is doubtful if it ever sailed. His design of turbine unfortunately was incapable of being made to produce large powers.

This great inventor's other interests ranged from electric machines and lighting to aerodynamics. Details of his inventions, of which there are several thousands, are all carefully written in his diaries which are preserved in the Technical Museum in Stockholm.

Gustav de Laval died at the age of 68 in Stockholm on 2 February 1913.

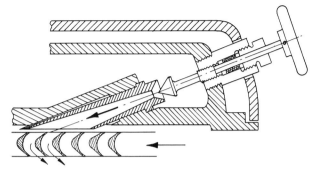

De Laval turbine nozzle

René Descartes 1596-1650

French philosopher and mathematician

Descartes was the founder of analytical geometry, the method of representing lines and curves by mathematical equations, thus making it possible to describe processes in space. Students of engineering and many other fields will be familiar with the plotting of graphs using 'Cartesian co-ordinates' but may not be fully aware of the fact that Descartes was an eminent philosopher, a great traveller and a soldier.

René Descartes, the latin form of his name being Renatus Cartesius, was born at La Haye, Touraine in France on 31 March 1596, the son of Joachim Descartes, a Counsellor in the Parliament of Rennes and member of a noble French family. He was educated at the Jesuit School of La Flèche and later visited Paris where he met Mydorge the mathematician and Mersenne the natural philosopher. In Paris he led a gay social life until 1614 when he settled down to a serious study of mathematics, but this was interrupted in May 1617 when he set off for the Netherlands to join the army of the Prince of Orange. In 1619 when the campaign had temporarily ceased, Descartes became bored through lack of activity and set off for Germany where he fought with the Bavarian army. It was in Bavaria that he received his famous 'philosophical conversion' as described in his work of 1672, *'Discours de la méthode.'*

He returned to France for a while and then travelled extensively in Italy and Switzerland. His attention turned from mathematics to science and, apart from another spell of soldiering at the Siege of La Rochelle in 1628, he studied, astronomy, meteorology, anatomy, and optics. In connection with optics he applied Snell's Law of refraction to the paths of

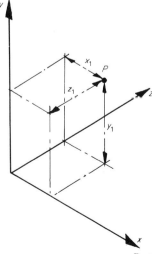

Rectangular Cartesian co-ordinates

light rays in raindrops and found, by calculating the paths of a thousand rays reflecting from the front and back of each drop, that a first and second bow would be formed. He also interested himself in the theory of music and wrote an essay *'Compendium Musicae.'*

To understand Descartes's science fully requires a study of his philosophy; he began by systematically doubting everything and then submitting ideas to his criterion of truth, their clearness and distinctness. The first reality of which he was certain was that of the thinking self, hence his famous statement, *'Cogito ergo sum'* (I think, therefore I am).

The influence of Descartes's philosophy was extensive and during the lifetime of Newton most scientists regarded themselves as followers of Descartes. There was a considerable amount of opposition to his philosophy, especially the part in which he attempted to prove the existence of God and, irritated by the attitudes of his critics, he left Holland, after twenty years residence, and went to Stockholm where he found favour in the court of Queen Christina. Within five months he contracted pneumonia and died on 11 February 1650.

Rudolf Christian Karl Diesel 1858-1913

German engineer

Rudolf Christian Karl Diesel, born in Paris on 18 March 1858, was the son of a Bavarian bookbinder.

In those days Paris was a technically-minded city abounding in first-class engineers and young Diesel was fascinated by the wonders of engineering which were to be found there. He was particularly attracted by Otto's famous gas engine shown at the World Fair of 1867. An extremely good pupil, he did especially well in engineering and draughtsmanship for which he was awarded a medal for outstanding academic ability. At the outbreak of the Franco-Prussian War the family, being German, had to move to London where Rudolf's father was able to find work. After a few weeks it was decided that Rudolf should attend a Technical school in Germany. So at the age of twelve he set off alone on the long and arduous journey to Ausburg to live with relations while he attended the Royal County Trade School. He left after three years, finishing top in the final year, and entered the Ausburg Industrial School to study Mechanical Engineering. At eighteen he attended the Technische Hochschule in Munich and passed the finals with the highest marks ever achieved to that date.

Professor Linde, one of Diesel's lecturers and the worldwide authority on heat engines and refrigeration, arranged for him to start work at the renowned Sulzer factory at Winterhur, Switzerland. Diesel began his practical training and, as was the custom for young engineers, became a proficient fitter and machinist. He very reluctantly left the shop floor to become manager of the Hirsch factory in Paris, but his disappointment was shortlived for he met a beautiful German girl called Martha Flasche with whom he fell in love. They married in 1883 and Martha who was very ambitious, was to have a considerable influence on the shy Rudolf.

At that time Diesel was experimenting with an ammonia engine and was contemplating the use of very high pressures in an air/liquid-fuel engine. His prototype of 1893 used such a high cylinder pressure that it was found necessary to inject the fuel at the end of the compression stroke to avoid premature ignition. The ability to obtain self-ignition, the main feature of the present day Diesel engine, was merely an additional feature not considered to be of any importance at the time. The prototype never ran satisfactorily; the high pressure involved caused the cylinder head to blow off. It took another four years of hard work before a reasonably reliable engine was produced. The new engine gave a spectacular performance developing 20 horse power with the amazing thermal efficiency of 26 per cent when running on

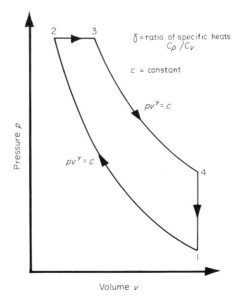

Diesel cycle

kerosene. Incidentally Diesel did not use coal dust in his first engines as popularly believed. With the minimum of development the new engine was put into production, and the work of Diesel was very soon accepted throughout the world and many of his engines were made under licence. At his wife's suggestion the modest Diesel reluctantly agreed to name the engine after himself.

However, Rudolf Diesel's enjoyment of such fame and fortune was soon to be marred by ill health, brought on probably by many exhausting legal battles over patent rights. He was rather naïve in business matters and lost a fortune through unwise financial speculations. His health deteriorated to such an extent that he had to enter a nursing home.

In 1913, his health having recovered, he travelled on the ship *'Dresden'* to England on business, which was to include a meeting with Sir Charles Parsons, when he disappeared. His neatly folded overcoat and hat were found on the afterdeck and a few days later his body was found by fishermen who returned it to the sea after removing objects of identification which were given to his son Eugene. In his cabin Eugene found his father's diary open at the day of his disappearance, but there was no suicide note, only a small cross in pencil beside the date.

Rudolf Diesel was a fascinating man with ideas too revolutionary for many of his contemporaries. He was a practical genius whose technical success was not to be fully enjoyed, partly through ill health and partly through bad luck with his business ventures. He suffered greatly from the frequent attacks on his patents made by unscrupulous men. Diesel is still remembered for his practical work on his engines in the laboratory and for his study of heat engine cycles which includes the 'constant pressure' or 'Diesel' cycle.

John Ericsson 1803-1889

Swedish engineer

John Ericsson ranks as one of Sweden's greatest engineers although most of his work was done in England and the United States of America. His name is to be found in most textbooks on thermodynamics in connection with the rather optimistic heat engine cycle known as the Ericsson Cycle.

He was born on 31 July 1803 in Warmland, Sweden. His forebears had been mining folk since the 17th century and his father was a qualified mining engineer. His brother Niels was a highly successful Naval Engineering Officer who built canals, locks, and docks in Stockholm, and was also the creator of the Swedish railway system.

The war with Russia, in which Sweden lost Finland, resulted in financial ruin for John's father who died four years later. Instead of going to University, John became a Naval Engineer like his brother and later transferred to the army as an Artillery Officer, where he learned a great deal about guns which was to be of great value to him later in life. He spent most of his spare time teaching himself engineering, mathematics, chemistry and English. He became obsessed with engines, in particular the hot-air engine of Stirling which he felt was bound to replace the steam engine.

In 1826 Ericsson travelled to England, in those days the Mecca of the world of engineering, with 1000 kronen in his pocket and a great deal of faith, but unfortunately, no one seemed to be interested in him or his engines. Ericsson, a man of great mental as well as physical strength was not easily daunted and, biding his time, went into partnership with a John Braithwaite from whom he learned a great deal about machine design. The firm built boilers, refrigerators, condensers and the first steam-operated fire engine to be used in London in 1828.

Ericsson had arrived in England at the very dawn of the railway age and had heard of the historic competition for steam locomotives called the 'Rainhill Trials'. He designed and built a locomotive which he called *Novelty*. It was beautifully made and extremely light in weight, and many regarded it as the favourite. Unfortunately for Ericsson, and perhaps fortunately for George Stephenson, a vital component failed and could not be replaced in time. Stephenson's *'Rocket'* of course won the race and the prize of £500, not to mention the fame that went with it. Undeterred by this temporary setback, Ericsson continued to bring out a major invention at the rate of about one a year including a rotary steam engine, a marine engine and a screw propeller which he patented in 1836.

Ericsson was still determined to produce a hot-air engine and was encouraged by the work of the Scottish inventor Stirling, who took out a patent for one in 1826. He tended to believe as many did at that time, that energy could be produced from almost nothing. A five horse-power hot-air engine which he built worked but had a disappointingly low efficiency. He turned next to the idea of screw propulsion for ships and approached the British Admiralty who were completely uninterested, so he decided to try his luck on the other side of the Atlantic.

In 1839 Ericsson sailed for the United States in the *'Great*

Western', but his wife, who was intensely jealous of his deep interest in engineering, did not wish to leave England and they separated. John was well received and the US Navy was far more sympathetic towards his views than the Royal Navy. He was asked to design the *'Princeton'*, a screw-propelled warship, which had all its machinery below the waterline, a distinct advantage over the traditional paddleship. The *'Princeton'* had a most novel design of 'high-speed' steam engine made by Ericsson. It was of a type known as a 'pendulum' engine with hinged rectangular pistons moving in segmental steamboxes. In common with many designers of steam engines, Ericsson had an unfounded fear of conventional pistons at high speeds.

Returning to his overwhelming desire to promote the hot-air engine, Ericsson built a large ship called the *'John Ericsson'* with a huge hot-air engine which was designed to develop 1000 horse power. The project was doomed to failure, the engine only managed to produce a mere 300 horse power and was even less efficient than a steam engine. After an early failure caused by the cylinders burning out, the engine was secretly replaced by a steam engine, but shortly afterwards the ship sank with all hands.

In spite of the tragic loss of the *'John Ericsson'*, Ericsson sold over 100 hot-air engines in two years, and altogether 3000 of his engines were installed in factories, docks, ships and mines. In the field of shipbuilding a great success was the *'Monitor'*, a semi-submerged warship with immense guns mounted in rotating turrets which was used to great effect against the South in the American Civil War; it was first offered to Napoleon III of France and cost 275 000 dollars. A further six gunboats were made at a cost of 10 million dollars and Ericsson also made torpedoes and guns up to 14-inch calibre for the North. These achievements were referred to in a famous dispatch from Ericsson to the president, Abraham Lincoln, "The time has come Mr President, when our cause will have to be sustained, not by numbers but by superior weapons . . . such is the inferiority of the Southern States in a mechanical point of view . . . that if you apply our mechanical resources . . . you can destroy the enemy without enlisting another man."

Thermodynamics textbooks often refer to the so-called 'Ericsson Cycle' but this is of academic interest only. It has an infinite number of stages of expansion and compression with intermediate reheating and intercooling respectively. Heating and cooling is at constant pressure, and a perfect heat exchanger is required. The efficiency is the same as that of the Carnot cycle, that is, the highest possible for a heat engine.

The latter part of John Ericsson's life was spent in exploring what today are called 'alternative energy resources' such as solar, tidal, wind and gravitational power. After a long life full of brilliant inventions and their applications, John Ericsson died quite suddenly at the grand age of 86 on 8 March 1889. A man of great strength, he had worked 12 to 14 hours a day, every day of the year for over 60 years until the very day of his death. He was buried with great honour in a special chapel at his birthplace in Warmland.

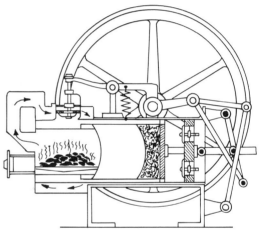

Ericsson's first engine

Leonhard Euler 1707-1783

Swiss mathematician

Leonhard Euler is regarded generally as the most prolific mathematician of all time. Some of his finest work was on the subjects of hydrodynamics and the elastic curves of loaded beams and columns. It is for this work that his name is familiar to the student of engineering. He was the first mathematician to formulate satisfactorily the laws governing the flow of fluids and to explain the importance of fluid pressure in relation to flow. The so-called Bernoulli Equation was first rigorously derived by Euler on the basis of a suggestion by his close friend, Daniel Bernoulli. As a mathematician, Euler was extremely interested in the geometrical form of deflection curves in Strengths of Materials and in 1757 he published an important paper on the buckling of struts. Most mechanical and civil engineers will be familiar with the 'Euler Strut Formula.'

Leonhard Euler was born at Basle, Switzerland in April 1707, the son of a Calvinist pastor, himself an amateur mathematician. Leonhard was educated at home by his father and then in 1723 he entered the University of Basle where he studied theology, oriental languages, physiology and mathematics. Initially he fully intended to follow in his father's footsteps and become a minister of religion but finally he chose a career in mathematics. After graduation, Euler must have shown a great flair for, at the age of 20, he

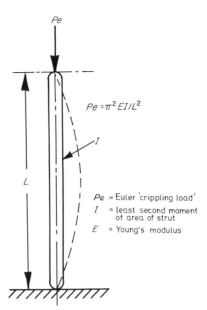

Euler strut formula

$Pe = \pi^2 EI/L^2$

Pe = Euler 'crippling load'
I = least second moment of area of strut
E = Young's modulus

was invited by Catherine the Great of Russia to settle in St Petersburg as a court mathematician. There he had his first meeting with the brilliant Daniel Bernoulli whom he later succeeded as Professor of Mathematics in St Petersburg in 1733. In 1741, Frederick the Great invited him to Berlin which is where he was to produce vast quantities of mathematical works over the next 25 years. The titles alone would fill a small book.

During 1735 in St Petersburg, Euler unfortunately lost the sight of an eye. Five years after his return to St Petersburg in 1766, he lost his sight completely, but this did not deter the great mathematician in any way; he dictated his mathematical works to two assistants and completed a total of 400 before his death.

In addition to producing many works on algebra and arithmetic, Euler was deeply interested in geometry and contributed a great deal to the theory of curves and surfaces, a field in which 'Euler's Constant' is well known. A well-known formula deals with the number of faces, edges and vertices of polyhedra. A keen astronomer, he studied the motion of the Moon in particular, and proposed an improvement on Kepler's theory of the Solar System. He also established the basic laws governing perturbation of heavenly bodies.

Leonhard Euler died in St Petersburg on 18 September 1783 at the age of 86.

Gabriel Daniel Fahrenheit 1686-1736

German instrument-maker and physicist

It was about 18 years before the publication of the famous memoir of Celsius that Fahrenheit, a German instrument maker published a paper on the thermometer. He was one of those physicists of the early 18th century whose pioneering work on the fascinating yet baffling problem of the nature of heat led famous men including, Lavoisier, Rumford, Joule, Helmholtz and Kelvin in their work. Several great names are associated with temperature scales, but the fame goes to Fahrenheit for the invention, first of the alcohol thermometer in 1709, then the mercury thermometer in 1714. Over two and a half centuries later this thermometer is still used throughout the world.

Gabriel Daniel Fahrenheit was born on 6 May, 1686 of German parentage in the city of Danzig, now Gdansk in Poland. He began his working life as a shopkeeper but apparently was not very successful and instead took up the study of physics. He moved to Holland, where he settled in 1717 to spend most of the remainder of his life, working as a glass-blower and a maker of precision meteorological instruments. After reading the works of the French physicist Amonton on the expansion of mercury with increased temperature and as the result of a meeting in 1708 with Ole Romer, he decided to try and make a reliable thermometer. The work is described in the florid style of those times in a paper in the 'Philisophical Transactions' of the Royal Society in 1724 (the original was in Latin). He wrote, "I was at once inflamed with a great desire to make for myself a thermometer so that I might, with my own eyes, perceive the beautiful phenomenon of nature," and then with great modesty, "I attempted to construct a thermometer but because of my lack of experience my efforts were in vain." Eventually he was successful for he continues, "When a thermometer was made (perhaps imperfect in many ways) the result answered my prayers and with great pleasure of mind I observed the truth of the thing."

Fahrenheit made many thermometers and investigated the freezing and boiling points of a number of liquids. He used many scales but finally settled on one which ranged from 32° to 96° where 32° is the temperature of melting ice and 96° the blood temperature of a healthy human. 0° corresponds to the temperature of a mixture of common salt and ice, the lowest temperature Fahrenheit could obtain. According to this scale water boils at 212° but it was not until after Fahrenheit's death that 212°C became the upper fixed point as in the modern scale. The scale which he produced has been in common use in English-speaking countries, although it has recently been superseded in Britain by the Celsius scale.

Daniel Fahrenheit made other contributions to science, discovering for example the important fact that water can remain liquid below its freezing point, a phenomenon known as 'supercooling'. He also found that the boiling point of water varies with air pressure and suggested, in a paper to the Royal Society in 1725, the use of this property as a basis for a barometer.

Fahrenheit died on 16 September 1736 at The Hague in Holland. He did not leave the world a great theory nor invent a wonderful machine. Nevertheless, his discovery of the humble thermometer was of the utmost value to science.

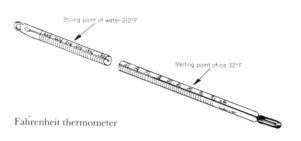

Fahrenheit thermometer

John Thomas Fanning 1837-1911

American civil engineer

John Thomas Fanning was born in Norwich, Connecticut on New Year's Day, 1837. He studied architecture and civil engineering and, having qualified, practised in both profession's until the outbreak of the Civil War in 1861. Fanning felt strongly about the Southern cause and at the age of 24 volunteered for service in the Confederate Army. By the cessation of hostilities he had risen to the rank of Lieutenant Colonel.

After the war in 1865, Fanning returned to civilian life and soon became a prominent engineer in the United States. He planned waterworks and water-power schemes in the northeast and west and was responsible for hydro-electric power and irrigation schemes on the Mississippi River at Memphis, the Spokane River in Idaho and others including the Missouri River. He also acted as a consultant engineer for a number of railway companies and was involved in the electrification of several railways in the United States.

Fanning was born about fifty years after the illustrious French civil engineer Darcy had published his well-known Formula for pipe-friction loss, but he also is remembered for a similar formula. The Fanning Equation differs only in the inclusion of a factor 4, so that values of the friction coefficient f have a quarter of the value of Darcy's. He wrote a number of very useful books on water supply and hydraulic power engineering, and contributed many papers to American Engineering Societies.

Fanning retired to St Anthony Falls on the Mississippi in Minnesota where he died on 6 February 1911.

Michael Faraday 1791-1867

British physicist
and engineer

Michael Faraday, a thirteen-year-old errand boy, was given the chance to learn the art of bookbinding by his employer, a London bookseller. He began reading some of the books he was binding, one of which was an encyclopaedia of electricity, and as a result he became deeply interested in science. He attended evening lectures on natural philosophy in Fleet Street for a shilling per lecture and made up his mind to involve himself in the world of science in any capacity no matter how humble it might be.

Such were the origins of Michael Faraday who was destined to be regarded by many as the greatest scientist of the 19th century and one of that select group of 'superscientists' which included, Archimedes, Galileo, Newton, Lavoisier and Darwin.

Faraday was born near London on 22 September 1791; his father, a blacksmith and his mother, a farmer's daughter were both from Yorkshire. He received only the bare rudiments of reading, writing and arithmetic at school and spent most of his time, by all accounts, playing marbles in the street. After completing his apprenticeship as a bookbinder and working as a journeyman Faraday was still hoping to become involved with science. He even wrote to the President of the Royal Society, the great Sir Humphrey Davy asking for employment, but did not even receive a reply. Later a relative who was acquainted with Sir Humphrey, had heard that the famous man had injured an eye in an experiment and required the aid of an assistant. Young Faraday's name was put forward and he was with some reluctance given the job at a wage of 25 shillings a week and the use of two rooms.

At first the young man was given little scientific work and spent most of his time rebinding books for the Royal Institution; however, when a laboratory assistant was sacked, the post went to Faraday. Davy gave him plenty of interesting work and also introduced him to many famous scientists. On travels abroad with his employer, Faraday met Ampère, Gay-Lussac, Clement and Désormes. For a young man who had never travelled more than a few miles, it was all an exciting adventure, he even saw Napoleon riding in his carriage.

After his return from the Continent, Faraday began a long period of research, making discoveries in chemistry, metallurgy and electricity. In 1817 he made his first great discovery, electromagnetic induction, and made his first transformer and dynamo. His concepts of lines and tubes of force and his work on their properties led to the formulation of the electromagnetic theory of light by Clerk Maxwell from which stemmed the invention of radio. Many of Faraday's discoveries were of vital importance to the electro-chemical and dyestuffs industries which were developing at that time.

In private life Michael Faraday was a very religious man and belonged to a stict sect known as the Sandemanians, named after their founder. His father was an elder of the sect which numbered only a few dozen. They followed simple Apostolic practises such as the ritual washing of feet, and regarded the saving of money as one of the most serious of sins. Faraday could easily have amassed a fortune from his many inventions but he never took out patents for his work and gave what little he did earn to charity. He probably did appreciate the monetary value of his discoveries however, for when the then Prime Minister, Gladstone asked him what use electricity was, Faraday replied, "Well, you will be able to tax it!" The story goes that he was asked the same question by an elderly lady, the reply on that occasion being "Madam, of what use is a new-born baby?"

Faraday had little interest in the opposite sex until he met a pretty girl called Sarah Barnard who was a member of the Sandemanians. In one of his many letters to her he wrote that his work, "amongst chlorides, oils, trials, Davy and steel," were follies and stupidity compared with his love for her. In 1820, when he was 29, he and Sarah married. She readily accepted the simple way of life which he preferred and became a devoted wife but they never had any children.

In 1857 Faraday was offered the Presidency of the Royal Society but he adamantly refused on the grounds that he did not accept its constitution and management; he also refused a knighthood. His last work, on the refraction of light by a magnetic field, was completed in 1862 when he was 71. Shortly after this, his mental and physical health deteriorated rapidly and he lived another five years in this tragic condition until his death in London on 25 August 1867. He was buried in Highgate cemetery.

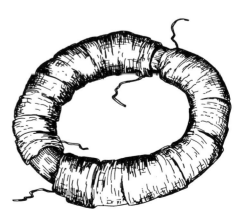

Faraday's first transformer

Michael Faraday, one of Britain's greatest scientists, made his many and important discoveries through hours of painstaking experiment. Like many such men of his day, he did not specialise but interested himself in a variety of scientific studies. He discovered liquified gases, made optical glass, invented the electric motor — the list goes on and on. This unselfish and modest man gave much to science and the world, with little thought of personal reward. He fully deserves the honour of having his great name immortalised in the S.I. unit of capacitance, the 'farad'.

Jean Baptiste Joseph Fourier 1768-1830

French mathematician

An unusual childhood, an adventurous manhood and a tragic death. This in brief was the life of Jean Baptiste Joseph Fourier one of the greatest mathematicians of all time and famous for his Fourier's Theorem and Fourier's Series.

Born at Auxerre in the Yonne Department of France on 21 March 1768, Fourier was the son of an impoverished tailor. At the age of eight he was orphaned and had little chance of receiving an education, but fortunately a former kind friend of his parents arranged and paid for his entry to the local Military School were he was later to teach. He was then sent, rather reluctantly, to the Abbey of St Bénoit-sur-Loire to study for the priesthood, but after two years he managed to leave and join the army where he hoped to become an artillery officer. Having played a modest yet rather dangerous role in the French Revolution, he was saved from the guillotine by the fall of Robespierre in 1794.

In 1795 Fourier accepted the position of Professor at the École Normale in Paris. Three years later he was with Napoleon's army of occupation in Egypt as a scientific adviser and at one time he was virtual governor of about half of the country. He returned to France in 1801 and was made Prefect of Isère, and it was about that time that he did his best work on the theory of heat conduction, for which he used his famous series to enable him to solve the law of heat propagation. He also formulated what is known as 'Fourier's Theorem' which states that any repetitive oscillation, no matter how complex, can be broken down into a series of pure sinewaves, each of constant amplitude, a process known as Fourier Analysis. The announcement of this theorem in 1807 brought him fame in the world of science and resulted in his title of Baron.

Fourier's fortunes fluctuated as France suffered changes of government. He published his book entitled *Analytical Theory of Heat*, which inspired Georg Simon Ohm to analogous reasoning with respect to the flow of electricity which eventually led to the familiar 'Ohm's Law'. The same work also greatly influenced the British scientists Lord Kelvin and Oliver Heaviside.

With the return of the Bourbons, Fourier received new honours and in 1862 he became a member of the French Académie, the following year succeeding Laplace as President of the Council of the École Polytechnique. Minor contributions to science by Fourier included the first use of dimensional analysis, an improvement on Newton's method for the approximate determination of the roots of an equation, and his treatment of statics in terms of the principle of virtual work.

Great men who have made tremendous contributions to science, must be permitted their little eccentricities. In the latter part of his life, Fourier, who had made such outstanding contributions to the study of heat flow, developed a theory that heat was essential to human survival and as a result he kept his home unbearably hot and swathed himself in many layers of heavy clothing. Such precautions were in vain and on 16 May 1830 Fourier tripped and fell to his death down the stairs of his Paris home.

Fourier's series:

$$\frac{a_0}{2} + \sum_{n=1}^{\infty}(a_n \cos nx + b_n \sin nx)$$

where $a_n = \frac{1}{\pi} \int_{-\pi}^{\pi} f(x) \cos nx \, dx,$

$b_n = \frac{1}{\pi} \int_{-\pi}^{\pi} f(x) \sin nx \, dx$

Benoit Fourneyron 1802-1867

French mathematician and engineer

Benoit Fourneyron was one of a group of talented hydraulics engineers, many of them his fellow countrymen, who developed the water turbine in the first half of the nineteenth century. He was born at St Etienne, France on 31 October 1802, the son of a mathematician, and attended the École des Mines there showing a great aptitude for mathematics. He went into the steel industry at Creuzot, Sâone-et-Loire, where he soon discovered the need for large powers to drive blowers and forge hammers in the production of steel.

Fourneyron's former professor, Claude Burdin, had written a memoir on a new type of hydraulic power machine as an alternative to the inefficient and bulky water-wheel, which dated back to the ancient Greeks. It had a vertical shaft and a runner with curved vanes, and inside these a ring of fixed vanes through which the water flowed radially outwards into the runner. This was completely different to the water wheel in which the water acts on only a few of the vanes at a time. It is interesting to note that Burdin was the first to use the word 'turbine'. His paper was given very little attention by the French scientific bodies but it did stimulate Fourneyron to carry out experiments with a 6 hp model.

At about the same time another Frenchman, Jean Victor Poncelet from Metz, had greatly improved the undershot water wheel by the introduction of curved vanes which greatly increased the efficiency from about 22 per cent to as much as 65 per cent. In 1826 he proposed an *inward* flow turbine but for some reason did not make one for another twelve years, after which it was improved by James Bicheno Francis the American engineer. Poncelet had followed a distinguished career in the army and after a few postings had been left for dead on the battlefield of Krasnoi in Russia. As a prisoner at Saratov he studied mathematics and on his release applied this knowledge to the subject of hydraulics.

The outward-flow turbine of Fourneyron is inherently unstable due to the diverging passages and is not very efficient at part load. Nevertheless at Besançon he developed a highly successful prototype of 50 hp with the encouragement of F. Caron, an ironmaster of Fraisans, and won a substantial prize from the Société d'Encouragement pour l'Industrie in 1833.

The Fourneyron turbine was very soon in demand all over the world and more than a hundred were built to operate at heads from less than one metre up to 120 metres with efficiencies of up to 75 per cent. In later years Boyden added a ring of fixed diffuser vanes at the runner outlet and raised the efficiency to 81 per cent, an amazing figure when one compares it with the 22 per cent for the straight-vaned water wheel. In 1895 Fourneyron turbines were installed in the world's first major hydro-electric scheme at the Niagara Falls, but the poor part-load performance eventually led to their being superseded by axial-flow Jonval turbines and later by Francis inward and mixed flow machines such as are used to the present day.

Benoit Fourneyron also experimented with steam turbines of the radial-flow type but with little success due to a lack of suitable materials and to poor workmanship, and it was left to Parsons, Rateau and De Laval to develop them. Later in life he became interested in politics and wrote articles on the state of the economy. In 1848 he sought election to the Constitutional Assembly. He died in Paris on 31 July 1867.

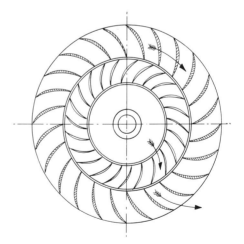

Fourneyron turbine

James Bicheno Francis 1815-1892

American hydraulics engineer

The tourist in Scotland who is interested in visiting some of the many hydroelectric schemes there, will find that three types of water turbine are used, the type depending upon the head of water available. Nestling in a valley by the roadside a small power station may be found using an impulse turbine, generally known as a Pelton Wheel. This utilises a very high head from some elevated reservoir and the water is supplied through a long pipeline or penstock. In dams at the end of large reservoirs two types of turbine are to be found. For low heads a propeller, or Kaplan turbine, is used and in the case of medium heads, an inward radial flow type known as the Francis turbine.

The latter is named after the distinguished American inventor and engineer James Bicheno Francis who made significant contributions to the development of hydraulic power in the USA during the 19th century.

Francis was born in 1815 in the hamlet of Southleigh in Devon, England, the son of John and Eliza (Bicheno) Francis. He received very little formal education but managed to get a job on the railways where he was able to acquire some knowledge of engineering. At the age of 18, seeing little prospects at home, he emigrated to America to seek fame and fortune, and in Connecticut, found work on the construction of the Stonington Railroad under the famous railway engineer G. W. Whistler. A group of ambitious young engineers in Lowell, Mass. operated under the highsounding name of 'The Proprietors of Locks and Canals of the Merrimac River,' and Francis joined them as a draughtsman, being promoted after four years to chief engineer at only 24 years of age. He also acted as a consultant to the hydraulic power industry and on the application of water power to cotton mills. It was at about this time that he married Sarah Brownell.

Francis carried out a considerable amount of research into the measurement of the flow of water in open channels and there is a well-known formula for the flow of fluids over weirs called the 'Francis Formula'. He also experimented with water turbines of high efficiency. His first design, in 1840, was under the patent of Howd of New York who had made a few rather crude wheels. Francis' turbine was of superior quality with a consequent great improvement in efficiency. In 1851 he carried out a series of exhaustive tests on two turbines, one an outward flow type or 'Fourneyron turbine', the other with an inward radial flow runner. Both were installed in mills in Lowell and were so successful that they drew a great deal of attention, especially in the case of the inward flow type to which his name has since been attached. Francis was responsible for formulating a design procedure for his turbine that is used to this day. In 1852 James Thomson, brother of Lord Kelvin, introduced the spiral inlet volute to the turbine and the system of adjustable guide vanes and flow control has survived with little change in modern machines.

A type of turbine similar to the Francis turbine is that invented by the Swedish engineer, Viktor Kaplan (1876-1934) in 1920, seventy years after Francis invented his turbine. It has a similar inlet volute and casing with adjustable inlet guide vanes. The runner however is in the form of an axial flow impeller with variable pitch blades to suit head and power variations. The Kaplan turbine is suitable for extremely low heads.

Francis had a wide range of interests, which included the preservation of timber, when he was the first to use the processes of 'burnettizing' and 'kyanizing'. He designed and developed methods of testing cast-iron girders, invented fire protection systems, and published over 200 technical papers, including work on the flow of water in pipes, channels, and over weirs.

This virtually unschooled village lad from Devon became one of the world's foremost hydraulics engineers, he contributed greatly to the development of engineering in the USA and was a founder member of the American Society of Civil Engineers. Francis was also instrumental in the development of Lowell from a small town to a great industrial centre. He died in Lowell on 18 September 1892 at the age of 77. About 11 years after his death, the Canadian Niagara Power Company installed a series of double Francis turbines at the famous falls which developed 10 250 hp. Modern machines are made in powers up to 20 000 kW with efficiencies of about 95 per cent.

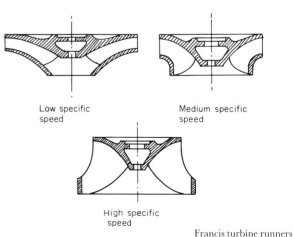

Francis turbine runners

William Froude 1810-1879

British hydrodynamicist

Students of Mechanical Engineering and Naval Architecture will almost certainly be familiar with the name Froude. It is used invariably when determining the resistance to motion of ships in connection with the 'Froude Number' a dimensionless quantity which is required in the design of ship's hulls.

William Froude (usually pronounced 'Frowd' but should be 'Frood'), was born at Dartington, a village on the River Dart in Devon, a part of the world closely connected with the sea and famous seafarers. He was the sixth son of Archdeacon Richard Hurrell Froude, rector of Dartington and Margaret Spedding a native of Cumberland. After an early private education, William spent seven years at Oriel College, Oxford, where he was taught mathematics by his elder brother Robert who was a leader of the 'Oxford Movement'. William was the only one of the six Froude brothers not to join the Roman Catholic Church.

When he left Oxford, Froude practised as a civil engineer for a number of years and came under the influence of the great bridge and steamship builder, I. K. Brunel, who encouraged him to take up naval architecture as an interest. Froude therefore decided to leave civil engineering at the age of 36 and, having sufficient means, worked privately on various aspects of ship hydrodynamics for scientific and personal reasons, as he was a keen yachtsman.

In 1856 Brunel asked Froude to investigate the behaviour of the Great Eastern, so Froude carried out a thorough investigation to determine the hull resistance and rolling characteristics on the ship and also on scale models. From this study he produced data which was of great value to naval architects for many years to come. His introduction of bilge keels enabled the rolling of vessels to be controlled and he also developed the technique of testing scale models of ships in towing tanks. When he applied to the British Admiralty for funds to build a large tank, he was strongly opposed by the leading members of the Institute of Naval Architects, led by the famous ship designer John Scott Russell who insisted that accurate data could be obtained only from towing-tests on actual ships. After a long and bitter struggle Froude was given the sum of £2000 with which he contructed a tank 250 feet long on his own land at Torquay. The tank consisted of an excavated trench spanned by a truck running on railway lines and towed by a steam winch. He was assisted in this venture by his son Robert Edmund Froude who, after the death of his father, built the Admiralty towing tank at Haslar, near Portsmouth.

Froude was not solely a theoretical man, he exhibited great manual skill not only in the construction of his ship models, but also in the making of instruments with which to measure the performance of the models. He introduced the use of paraffin wax models and waterline cutting machines, and among the many instruments which he made were towing resistance recorders, governors, roll indicators and propeller-engine dynamometers. The hydraulic dynamometer designed by Froude had immense advantages over the friction brake. The Heenan-Froude dynamometer is found in many engineering laboratories and industrial testbeds to this day. The improved modern version of this instrument is by Osborne Reynolds.

The resistance of a ship to motion is a combination of two parts, the wavemaking resistance, also called the 'residual drag' (where the energy is lost in wave-making) and the viscous or skin resistance caused by drag on the submerged surface of the hull. These two components scale differently between a model and a full-scale ship, so that separate investigations have to be undertaken. Wavemaking resistance is determined by towing tests, and skin resistance by the drag on flat submerged surfaces. Froude showed that for a true comparison between towing tests on models and tests on full-scale ships the velocity must be proportional to the square root of the length. The dimensionless quantity known as the 'Froude Number' is given by $Fr = c/\sqrt{(Lg)}$, where c is velocity, L is length and g is the acceleration due to gravity. The skin resistance was determined by Froude by towing planks of different sizes and surface roughness in Torbay or in his test tank.

During his lifetime William Froude contributed many excellent articles on ship hydrodynamics to the Transactions of the Institute of Naval Architects and to the British Association. He was made a Fellow of the Royal Society in

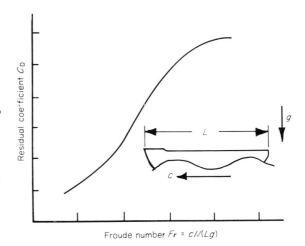

Froude number $Fr = c/\sqrt{(Lg)}$

Residual drag for ship

Luigi Galvani 1737-1798

Italian anatomist

1870 and he can safely be called the 'father of ship hydrodynamics'. Unlike many great engineers he came from a comfortably-off family and took up engineering more or less as a hobby. He was a tireless experimenter and contributed a vast amount of invaluable information to the science of ship design. While holidaying in South Africa, William Froude contracted dysentery and died at Simonstown, Cape Province on 4 May 1879.

The name of the 18th century anatomist, Luigi Galvani, has become firmly entrenched in the English language where a number of words in common use are based on his name. 'Galvanizing' refers to the process of electroplating iron with zinc, (the word is used also for other methods of zinc coating) and a number of instruments are named after him, for example the galvanometer and the galvanoscope. Yet the fame of this Italian doctor of medicine is based upon a complete misconception concerning a phenomenon which he termed 'animal electricity'. Nevertheless his devoted and painstaking investigation of a completely unforseen natural process led to the explanation of Alessandro Volta which opened up a new age of electrical science.

Luigi Galvani was born at Bologna in northern Italy on 9 September 1737 and after first studying theology he turned to medicine, in which he obtained a degree at Bologna University. It was also at this time that he married the daughter of his former guardian and teacher. After carrying out some experiments on the ears and kidneys of fowls, he was appointed lecturer in medicine at the University in 1762. In 1775 he became Professor of Anatomy and later also Professor of Obstetrics as a result of a brilliant treatise on the semi-circular canals in birds.

He was deeply interested in the nervous systems of animals and did a great deal of research on the mechanical and electrical stimulation of nerves. He made the interesting discovery that the muscles of frog's legs could be made to respond to the discharge from an electrical discharge machine when situated some distance away. The only possible source of energy must have been electromagnetic radiation, that is radio waves! Here was an example of electromagnetic induction forty years before its discovery by Faraday.

Galvani also proved that these muscular convulsions could be produced by lightning or under troubled atmospheric conditions. Later he found that frog's legs being dried out of doors for future dissection experienced similar convulsions when attached to iron railings by non-ferrous metal skewers. When this effect was repeated indoors away from any possible source of electricity, the same phenomenon took place if the frog's legs formed part of an electrical circuit which included one or more pieces of metal. He believed, not unnaturally, that the source of electricity was inherent in the muscles or nerves which acted rather like a Leyden Jar. Galvani spent eleven years on these experiments and first announced them in a paper, *'De viribus electricitatis in motu musculari commentarius'* published in 1791 when he was 54 years old.

Alessandro Volta investigated Galvani's work and showed that electricity could be produced by dissimilar metals in the complete absence of a frog's leg or any other animal material. A great controversy was to follow between supporters of Galvani's theory and those who adhered to Volta's discovery. The argument was finally settled in favour of Volta in 1800, two years after the death of Galvani, when he made his famous battery or 'Voltaic Pile'.

In 1796 Napoleon Bonaparte invaded northern Italy and set up the Transalpine Republic which included Bologna, formerly a Papal State. Luigi Galvani refused to swear the oath of allegiance to the new constitution and was dismissed from all his offices. Friends managed to arrange for his reinstatement in 1799, but due to domestic bereavement, poverty and severe criticism of his work his health deteriorated and he died on 4 December 1798 at Bologna.

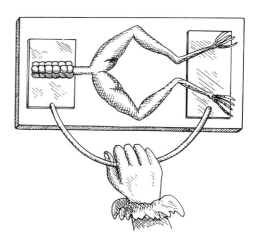

Galvani's frog's legs demonstration

Incidentally his rival Volta was one of a delegation sent to welcome the victorious Napoleon Bonaparte.

Luigi Galvani was a born investigator of the wonders of nature, who devoted himself entirely to a line of research which was regarded generally as being of little importance and which had no prospects of academic or financial reward. It often happens that from such modest beginnings great scientific advances are made.

Karl Friedrich Gauss 1777-1855

German mathematician, astronomer and physicist

Karl Gauss, son of a poor gardener, was born at Brunswick, Germany on 30 April 1777. He was an extraordinarily intelligent child who grew up to become a mathematician, astronomer and physicist of the highest order. In many ways he was comparable with Archimedes or Newton but he had a far wider range of ability than either. His talent for complicated mental computation was most unusual even for mathematicians of his calibre.

The name Gauss was associated formerly by physicists and electrical engineers with the cgs unit of force in a magnetic field but, with the advent of SI units, Gauss lost that honour to the Yugoslavian engineer, Nikola Tesla, the unit now being known as the 'tesla'. His name however is given still to the process called 'degaussing', a method used in wartime to protect ships from magnetic mines. The instrument used for measuring magnetic field strength is known as a 'gaussmeter' and the statistician is familiar with the Gaussian (or Normal) Distribution Curve.

The young Karl Gauss was given financial assistance to attend a good school in Brunswick where by his early twenties he had mastered the works of great men such as Newton, Euler and Lagrange. At Göttingen University under the patronage of the Duke of Brunswick he compiled his most famous work, *Disquisitiones Arithmeticae* which was published in 1801. After the death of his patron, Gauss was made the Director of Göttingen Observatory and became responsible for the establishment of the geomagnetic observatory there. He was also made adviser to the Hanoverian and Danish Governments in a geodetic survey.

In 1809 Gauss published another important paper on the motions of celestial bodies. This contained outstanding calculations on the orbits of planets and comets and the prediction of the point of reappearance of the newly discovered minor planet Ceres.

Many of his greatest mathematical papers were not published until well after his death. These were found to contain theories which were rediscovered and published by others, which was rather embarrassing for a number of them, who had become famous as a result of what they thought to be original work. Gauss was fond of using the definitions of electrostatic and electromagnetic mathematical definitions in connection with other applications, as for example in the 'Theory of Knots' which is only now becoming of practical value.

Gauss remained at Göttingen working hard until his death on 23 February 1855 at the age of 78.

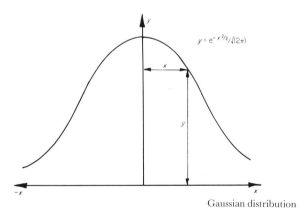

Gaussian distribution

Joseph Henry 1797-1878

American mathematician and electrophysicist

Although many engineering students are familiar with the unit of self-inductance or 'inductance' as it is usually called, they probably know nothing about the man after whom the SI unit, the 'henry' is named. It is claimed that Joseph Henry discovered electromagnetic induction independently of the much more famous Michael Faraday.

Joseph Henry, born at Albany in the State of New York on 17 December 1797, had the barest of primary and even secondary education, but he was an avid reader. Like his contemporary, Faraday, a chance encounter with a book on science led to his serious studying to gain entry to the Albany Academy where he studied chemistry, anatomy and physiology with the intention of becoming a doctor of medicine. He eventually decided, in 1825, to give up medicine in favour of engineering. After qualifying, he worked at the Albany Academy for some years as a teacher and was appointed Professor of Mathematics in 1826. Six years later he was made Professor of Natural Philosophy at the prestigious University of Princeton, where he taught a wide range of subjects–mathematics, chemistry, physics, mineralogy, geology and, later on, astronomy and architecture.

From 1846 and for the remainder of his long life, Henry was the first director of the Smithsonian Institute in Washington. This well-known organisation was founded and endowed by James Smithson, the very wealthy natural son of the Duke of Northumberland, for the purpose of the increase and diffusion of knowledge among men. From the Smithsonian many other scientific bodies in the United States originated, including the Meteorological Bureau and the National Museum. Joseph Henry was also a founder member of the National Academy of Science and was active in the establishment of the American Association for the Advancement of Science in 1848. During the Civil War, Henry was one of Abraham Lincoln's technical advisers.

From his early days in Albany, Henry was deeply interested in electromagnetism, he greatly improved the electromagnet by using insulated wire, rather than insulating the core, so that he could use many hundreds of overlapping turns. He made a huge electromagnet for Yale College which could raise a load of one ton, a record at that time. His discovery of self-inductance was of the utmost importance in the development of electrical engineering, for he showed that when a current flowing in a wire or coil changes in strength with time, a second weaker current of opposite sign is produced which retards the change in the original current.

Another important discovery was that a varying current flowing in a coil produced a current in an adjacent coil, the basis of the transformer. Henry even managed to induce a current in a coil situated at quite a distance from the energisd coil and was able to magnetise a needle by means of a lightning flash eight miles away; this was possibly the first use of radio waves. In 1835 Henry made the first relay and invented non-inductive windings.

The American inventor was mortified when Faraday, who published his work on induction in 1831, was given all the credit; Henry's other duties of teaching and his involvement on numerous committees had delayed the publication of his work. He wrote in one paper, "I have made several other experiments in relation to electricity, but which more important duties will not permit me to verify in time for this paper." He was given some recognition and was honoured by having his name given to the unit of inductance in 1893. The 'henry' is now the SI unit.

Henry also paralleled Faraday by constructing a small electromagnetic motor which he regarded as 'merely a philosophical toy!' In 1831 he successfully designed and operated the *first* electric telegraph over a distance of a mile, and later used it in the Meteorological Bureau to transmit data for weather forecasts obtained from a large corps of volunteer observers which he had organised. With the help of his brother-in-law, Henry carried out investigations into sunspots and solar radiation. Using a sensitive thermo-galvanometer he showed that sunspots radiated less heat than the normal surface of the sun. A very important discovery of this much undervalued scientist was that the discharge of a Leyden jar, an early form of capacitor, was oscillatory, the oscillations being of a very high frequency. This was of great value to Hertz in his later work on electromagnetic radiation.

Joseph Henry died on 13 May 1878 in Washington DC. He is generally considered to be the greatest American scientist with the possible exception of Benjamin Franklin. Many claim that he actually beat the English physicist Michael Faraday by at least one year in discovering electromagnetic induction and also was the first to make an electric motor. It is certain that had he devoted his great genius more exclusively to research instead of administrative duties, Henry would have obtained far more recognition and been of greater service to American science.

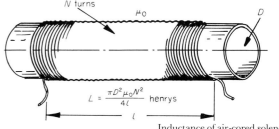

Inductance of air-cored solenoid

$$L = \frac{\pi D^2 \mu_0 N^2}{4l} \text{ henrys}$$

Hero of Alexandria c. A.D. 60

Alexandrian mathematician and engineer

Many textbooks on Engineering Thermodynamics refer to the famous reaction steam turbine of Hero (or Heron) of Alexandria at the beginning of the chapter on steam turbines, and many give a highly decorative illustration of this ancient and wonderful machine. It is remarkable that it took another 18 centuries before practical turbines were made by men like de Laval, Rateau and Sir Charles Parsons.

Very little is known of the private life of Hero but much has been written about his many ingenious inventions and of his four famous books on mathematics and physics, namely, *Metrica, Mechanics, On The Dioptra,* and *Pneumatics*. *Metrica* deals with the measurement of plane and solid geometrical figures, and features conic sections, frustum of a cone, and the five regular or Platonic solids. There is a formula for determining the area of a triangle from the lengths of the sides and a method for calculating the square root of a non-square number.

Hero's *Mechancis*, the only existing copy of which is in Arabic, contains the parallelogram of velocity, the laws of levers (both straight and cranked), the inclined plane, and much useful information on gears and positions of the centre of gravity of bodies.

On The Dioptra describes various instruments, for example, a type of theodolite, for which Hero used a refined screw-cutting technique, its use in the boring of tunnels is discussed, and the work also includes a description of a hodometer for measuring the distance travelled by a wheel.

Hero's *Pneumatics* contains all manner of interesting mechanical devices, many of them centuries before their time, and powered by atmospheric pressure, heated air or gas, water or steam pressure. He discusses siphons, fountains, pumps, water and wind-powered pipe organs, and automata in the form of humans and animals. Many of these weird and wonderful machines were used in the temples to create magical effects. A handful of powder sprinkled on to an altar fire would produce hot air which, on displacing water would mysteriously open the temple doors. He even invented the first 'coin-in-the-slot' machine, a holy water dispenser described by the rather mercenary priests as a 'sacrificial vessel which flows only when money is introduced'.

Another 'first' was Hero's 'taximeter' fitted to passenger chariots to calculate the fare. His best known invention, the steam turbine, or 'aelopile' as it was known, consisted of a spherical rotating boiler fitted with two tangential nozzles working in a similar manner to the familiar garden lawn sprinkler. Although it is doubtful if any appreciable power was developed, it is interesting to note that it was Newton who first explained the meaning of 'reaction' more than 16 centuries later.

The principle of the aelopile was used by the steamboat pioneer Oliver Evans of Philadelphia (1755-1819). His turbine used steam at 56 pounds per square inch and ran at speeds between 700 and 1000 rev/min, producing 'the power of two men?' Richard Trevethick also considered its possible use for marine propulsion.

There is no doubt that Hero, and probably many of his contemporaries, foresaw a great number of the important principles of physics which were not to be reintroduced for many centuries. Hero for example wrote on the compressibility and density of air and assumed it to be composed of small particles free to move relative to one another. This was some 1500 years before Boyle.

Hero's steam turbine

Heinrich Rudolf Hertz 1857-1894

German physicist

Heinrich Rudolf Hertz is one scientist whose name has become more prominent by the introduction of the new Système International d'Unités. Previously the term for frequency of periodic phenomena in physics was unconnected with any person's name; however, it is now named in honour of the German physicist who first demonstrated the transmission of electromagnetic waves and thus paved the way for the invention of wireless telegraphy. Frequency was formerly given a variety of names, varying from 'cycles per second' to 'vibrations per minute'. Frequency is now measured in cycles per second and called 'hertz'.

Heinrich Rudolf Hertz, son of a highly successful lawyer, was born in the city of Hamburg, Germany, on 22 February 1857. At school he was particularly fascinated by experiments in optics and mechanics; he also attended additional evening classes on engineering and the use of measuring instruments. On leaving school young Rudolf decided to be an engineer and attended the Polytechnic Institute at Munich, but after a year's study, he realised that his future lay in the field of pure science and he moved to Berlin University to study under that famous trio, Kirchhoff, Bunsen, and Helmholtz. He later went to Kiel for further study and was then appointed Professor of Physics at Karlsruhe where he was to carry out his historic experiments on radio waves.

About 23 years before Hertz carried out his classic experiment, Clerk Maxwell, the great Scottish physicist had put forward the theory that electromagnetic waves generated by electrical oscillations could exist in the aether and that they travelled at the speed of light. Hertz used a spark-coil to produce the electromagnetic radiation with large plates connected to each side of the spark-gap, and a receiver consisting of a wire hoop with a small gap. When the receiver was positioned near the transmitter a feeble spark was observed in the receiver gap. Hertz went on to show that the radiation, which had a wavelength of only 24 cm, behaved in the same manner as light and could be reflected, refracted and polarised. Moreover he showed that the velocity of the radiation was the same as that of light, approximately 30 000 000 m/s.

Sir J.J. Thomson described the experiments of Hertz as "one of the most marvellous triumphs of experimental skill and ingenuity in the whole history of physics." Sir Oliver Lodge said, "Hertz stepped in before the English physicists and brilliantly carried off the prize."

Although Hertz' investigations in the field of elasticity formed a relatively small part of his scientific achievements, they were nevertheless of vital importance to the engineer. This work included an analysis of the stresses and deflections of plates on elastic foundations and the compressive stress distribution of elastic bodies in contact. The latter deals with what are referred to as 'Hertzian Stresses' and is of great importance in the design of ball and roller bearings.

In 1889 Hertz succeeded the great thermodynamicist Clausius (of entropy fame) as Professor at Bonn but his brilliant career was short lived. Five years later at the age of 37, he died as a result of blood-poisoning on New Year's Day 1894. Seven years later the Italian engineer, Guglielmo Marconi successfully transmitted a wireless signal across the Atlantic Ocean.

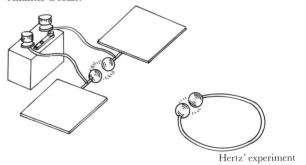

Hertz' experiment

Robert Hooke 1635-1703

British physicist

One of the most brilliant and versatile of 17th century scientists, Robert Hooke has been known to generations of school pupils for 'Hooke's Law' and to mechanical engineers for the universal coupling known as 'Hooke's Joint'.

Robert Hooke was born at Freshwater on the Isle of Wight on 18 July 1635, and was by all accounts a sickly child, not attending school until he was 13 years old. He spent his time at home drawing and playing with mechanical toys of his own

construction until he was old enough to go to Westminster School in London. Later he obtained a place as a chorister at Christ Church, Oxford where he came into contact with many famous scientists. Being deeply interested in science and also being a highly skilled mechanic, he was able to help some of them in their research work. He became an assistant of the famous Robert Boyle and was of great assistance in helping him to construct the air pump which Boyle used for his historic gas experiments. Hooke began to carry out scientific investigations of his own on pendulums and springs for clocks and when the Royal Society was founded in 1662, Boyle successfully proposed Hooke for the newly created post of Curator. As Curator, Hooke was responsible for the setting up of three or four interesting experiments of a suitable kind for each meeting. He later became Secretary of the Royal Society.

Robert Hooke was fascinated by the nature of light and in 1665, in his work *Micrographia*, he suggested a wave theory of light and described experiments on the colour of light passing through membranes. Twenty years before Newton began his work on *Principia*, Hooke proposed a theory of planetary motion based upon the balance between centripetal force and gravitational force and actually suggested to Newton that the gravitational force between the earth and the sun was inversely proportional to the distance between them!

After the Great Fire of London in September 1666, Robert Hooke was one of several famous men involved in the rebuilding of the city. His model of London showing his ideas for rebuilding so impressed the magistrates that they immediately appointed him as a surveyor. He assisted in the rebuilding and actually designed several buildings.

Among Hooke's many inventions were: spring control of the balance wheel of a watch, the compound microscope, a wheel barometer, the universal joint (although this is also accredited to Cardan), the spring balance and the reflecting telescope later referred to as the Newtonian telescope. In 1676 he experimented on the elastic properties of materials, particularly metals, and in 1678 produced his famous paper *'De Potentia Restitutiva'*, which embodied the principle now known as 'Hooke's Law.' This states that for an elastic material, the extension of a rod, spring, etc. subjected to a load, is proportional to the magnitude of the load. It is upon this apparently simple law that the whole study of the mechanics of elastic bodies is founded. It is interesting to refer to Hooke's own words, ". . . or take a wire 20 or 30 or 40 feet long and fasten the upper part to a nail and the other end to a scale to receive the weights . . . measure the distance of the scale to the ground and set down the distance . . . put weights on the scale and measure and compare the several stretchings and you will find them always being the same proportions one to another that the weights do that made them."

Hooke experimented with the transmission of sound waves in various materials including a taut metal wire, and wrote "a whisper could be heard over a furlong," and that, "an improvement of ten times that distance is possible." He also interested himself in flying machines and believed in the 'fixed wing' solution.

It is rather unfortunate that such a fine and prolific scientist should have had his reputation suffer as a result of his bitter struggles with other scientists such as Newton and Huygens. Although Hooke's portrait was certainly painted at least once during his lifetime, no copy appears to exist now, not even at the Royal Society. The portrait shown is based on the illustration in *Isis*, **33,** 15-17 (1941).

In 1696 Hooke's health deteriorated rapidly and after seven years of illness he died in London on 3 March 1703.

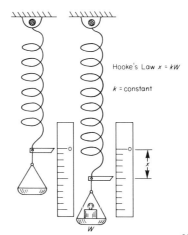

Hooke's Law

James Prescott Joule 1818-1889

British scientist and engineer

According to the Oxford English Dictionary the 'joule' may be pronounced either 'jowl' or 'jool'. The surname of the man after whom it is named however is certainly pronounced 'Jowl'. Students and, alas, many lecturers are under the impression that Joule was French, or from a French-speaking country, and may even be tempted to say 'Zhool'.

James Prescott Joule was actually born at Salford near Manchester, England on 24 December 1818, the second of the five children of Benjamin Joule, a successful brewer whose ancestors had been Derbyshire farmers for generations. James was apparently a very delicate boy for he suffered from a spinal ailment which left him with a slight permanent limp. In later life he was afflicted with frequent nose bleeding, possibly haemophilia.

As a child James and his elder brother Benjamin were taught by private tutors. James was fascinated by scientific toys and amused himself by giving servants and relations electric shocks. He also flew kites in thunderstorms in the manner of Benjamin Franklin. One of his tutors was the ageing John Dalton, the Father of Modern Chemistry who had been reduced to earning a little pocket money by teaching the children of the well-to-do. Apparently the tuition was not satisfactory, at least from the father's point of view, but the great man Dalton made a lasting impression upon the boy.

At the age of fifteen James began work in his father's brewery where he became familiar with both engineering and chemistry. In those days a brewery was one of the few places where it was possible to find chemical processes such as fermentation taking place, or machinery such as engines and pumps handling large quantities of liquids and gases. When he was twenty, Joule set up a laboratory in his father's house and was soon attending meetings of the Manchester Literary and Philosophical Society where he met many famous scientists and engineers. He thrived in the strong technological atmosphere of the industrial provincial city and would probably never have formulated his principle of the conservation of energy if he had been trained in the conventional manner of the Universities which, in those days, were dominated by Newtonian mechanics.

Joule is of course best known for his determination of the mechanical equivalent of heat for which he received many honours in his day, and in recent years has had his name adopted as the SI unit of energy, the 'joule' (J). Formerly 'J' was used for the energy conversion factor so that it was equal to approximately 778 foot-pounds per British Thermal Unit. In conjunction with Lord Kelvin (William Thompson), Joule discovered the 'Joule-Thompson Effect', that is, the slight cooling effect in a gas when expanded through a porous plug. This discovery was to be the basis of the liquid-gas and gas refrigeration industry.

In his twenties Joule experimented on the power to drive electric motors and the heating effect of an electric current. He began his work on energy conversion using a dynamo, then by measurements on the heating of water flowing through a pipe by an electric current. He also carried out his famous experiments using a 'paddle wheel' in water. One can well imagine the bearded respectable Victorian delivering his lecture on his Law of the Conservation of Energy to his equally respectable fellow members of the 'Lit. and Phil.' in the St Anne's Reading Room in Manchester in May 1847. The lecture was titled 'On Matter, Living Force and Heat'. His figure at that time for the mechanical equivalent of heat was 781.8 ft-lb to raise the temperature of 1 lb of water through 1°F.

In his memoir 'On the Mechanical Equivalent of Heat' which Joule read to the Royal Society in 1849, he says, "In accordance with the pledge I gave the Royal Society some years ago, I have now the honour to present it with the results of the experiments I have made in order to determine the mechanical equivalent of heat with exactness."

Later in life Joule experimented with electric welding and also made the first mercury displacement vacuum pump. He proposed the Constant Pressure (Joule) cycle for hot air engines which he said could be run at much higher temperatures than the steam engine and thus attain higher efficiencies. The same cycle, under the name 'Brayton Cycle' is used for gas turbines.

In 1847, the year of his historic lecture in Manchester, Joule at the age of 29 married Amelia Grimes, the daughter of the Comptroller of Customs at Liverpool. She died tragically after only seven years of marriage leaving a son and daughter. Joule was terribly shaken by her death and although he continued to experiment in his laboratory at home, he became increasingly retiring. He remained aloof

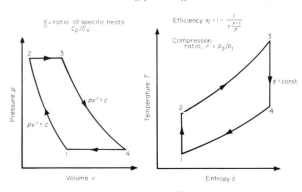

Joule constant-pressure cycle

from the Royal Society of which he became a member in 1847, and adamantly refused to take up a University post. After a long period of ill health lasting 17 years, James Prescott Joule died at the age of 78 on 11 October 1889 at 12 Wardle Street in Sale, Cheshire. Lord Kelvin had said of him in 1882, "His boldness in making such large conclusions from such small observational effects is almost as noteworthy and admirable as his skill in extorting accuracy from them."

Théodore von Kármán 1881-1963

Hungarian/German aerodynamicist

In the study of fluid mechanics, the budding engineer is quite likely to come across the name of von Kármán in connection with pipe friction, boundary layer theory and the production of vortices. Born in Hungary, von Kármán worked in Germany until his late twenties then went to the United States where he helped to found the aircraft and rocket industry.

Théodore (Todor) von Kármán was born in Budapest on 11 May 1881, the son of a University Professor, Maurice (Mor) Kármán who had been knighted by the Emperor Franz Joseph of Austria-Hungary for reorganising secondary education in Hungary. Von Kármán was given his primary education at the German Gymnasium, then he attended Budapest Technical University where he became deeply interested in problems in applied mechanics. After spending a year on compulsory service in the Artillery, he worked for three years as an instructor in the University.

In 1906 he was employed at the University of Göttingen working with the famous Ludwig Prandtl, rightly called 'the father of aerodynamics', on problems connected with aerodynamics, the variation of the specific heat of gases with temperature, and turbulent flow over various shaped bodies. It was when investigating the flow of air over a cylinder that Kármán observed the regular trail of vortices which has since become known as the 'Kármán Vortex Street', or in German 'Kármánsche Wirbelstrasse', and he derived an expression for the frequency with which the vortices were shed. After the dramatic failure of the Tacoma Narrows Bridge in the United States, in a 42 mile/h wind, he proved that the oscillations which caused failure were excited by the vortex shedding frequency.

During World War I, Kármán worked for the German air force on propeller design, which included the application of contra-rotating propellers, the synchronisation of aircraft machine guns and the design of helicopters. After the war he contributed greatly to the aerodynamic design of gliders, little knowing that he was laying the foundations for the Luftwaffe in World War II for the Nazis, of whom he strongly disapproved. At the University in Göttingen, Kármán studied turbulent flow in friendly rivalry with Prandtl and investigated the aerodynamic drag on the surfaces of aircraft and rockets.

In 1926 Theodore von Kármán left Germany for a tour of the USA and Japan. In the USA he was greatly interested in the centre at the University of Technology in California endowed by the philanthropist Guggenheimer. He returned there permanently in 1929 as Director of the Aeronautical Laboratory where he helped to found the Aerojet Engineering Corporation and institute the Jet Propulsion Laboratory where rocket research was carried out.

Von Kármán never married; his sister Pipö looked after him at his home in Pasadena, California until her death. He died on 7 May 1963 at the home of a friend in Aachen while holidaying in Germany. This great man of aerodynamics was honoured in the United States by being awarded the US Medal for Merit, the Franklin Gold Medal and the National Medal for Science. He also received honours from many other countries.

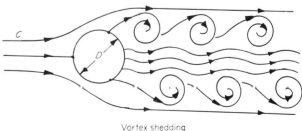

Vortex shedding frequency $f = 0.207\, C/D$

Kármán vortex street

William Thomson, Lord Kelvin 1824-1907

British mathematician, scientist, engineer

William Thomson was born in Belfast, Northern Ireland, in 1824, the son of a self-educated mathematics teacher, who later became Professor of Mathematics at Belfast and then, when the family moved to Scotland, at Glasgow University. He personally undertook the education of his sons William and James and must have made a tremendous success of it judging by the results. William managed to matriculate into Glasgow University at the incredibly early age of ten.

William was only six years old when his mother died and the Professor acted as both father and mother to the children at their University lodgings. They lived very close to the most appalling Glasgow slums where on a Saturday night the air was full of the coarse cries of drunken revellers and impending murder.

William Thomson was an outstanding student at Glasgow University and had produced no less than twelve research papers before he had even graduated. At the very early age of twenty-two he became Professor of Natural Philosophy at Glasgow and after a further year his work was known throughout Europe. Such was the apparently inexhaustible flood of research works which he produced that he rarely had time to read the work of contemporaries and preferred to do his research without dependence on others.

His finances on the other hand were more disciplined and unlike most scientists he made a considerable fortune out of his inventions. During his lifetime he took out 70 patents and amassed the sum of £160 000, a fortune in those days. The success of the first Atlantic Telegraph Cable was due largely to William Thomson's engineering genius. He determined the most suitable system for transmitting the signals, and invented instruments of the highest quality for sending and receiving the signals. This achievement made him publicly famous and had a tremendous effect on the development of precision electrical instruments and the accurate determination of electrical and physical units which led to the founding of the National Physical Laboratory. The following are just a few of the many instruments which he invented; the Kelvin ampère-balance, the Kelvin bridge for the accurate measurement of low resistance, the Kelvin electrometer which is the same as the quadrant electrometer, and the Kelvin-Varley slide, a constant resistance voltage divider. In the field of marine instruments he invented the Kelvin sounder and the Kelvin compass. He discovered the 'Kelvin effect' whereby alternating currents, especially of high frequency, tend to concentrate at the surface of a conductor, a phenomenon of great importance in radio and high-frequency heating. Kelvin's Law is concerned with the most economical size of an electrical conductor.

Thomson was also greatly interested in the theory of the conservation of energy and the mechanical equivalent of heat, a field in which he collaborated closely with James Prescott Joule, as for example, in the Joule-Thomson Effect. He developed the idea of an absolute scale of temperature, of vital importance to the study of thermodynamics, in which the interval is the 'kelvin' symbol K, named in his honour and equal to 1° Celsius, so that $K = °C + 273$. He also explained the use of the Carnot cycle in relation to the heat pump and the refrigerator.

In 1852, when he was 28, Thomson married Margaret Crum, a handsome, cultivated but unfortunately delicate woman, who was taken ill on their honeymoon and continued in ill health for the remaining 18 years of her life. Thomson was knighted in 1866 for his many contributions to science and engineering and took the title of Lord Kelvin of Largs. In 1896 he celebrated his fifty years as Professor at Glasgow University and in his Jubilee speech surprisingly described his life as a 'failure', that is, "A failure to fit physical science into the engineer's concept of nature."

William Thomson, Lord Kelvin died on 17 December 1907 in Netherhall near Largs in Ayrshire, Scotland. He was 83 and had preserved his wonderful clarity of mind until the end of his days. This man who had described himself as a 'failure' produced 600 scientific papers, took out 70 patents for inventions, received honours bestowed by 250 academies and societies and made a considerable fortune.

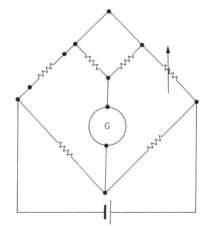

Kelvin bridge (Thomson double bridge)

Gustav Robert Kirchhoff 1824-1887

German physicist

Kirchhoff's first research work was on the conduction of electricity in wires and it is for this mainly that he is known to the engineer. His laws for closed electrical networks namely that the algebraic sum of the currents at a junction is zero and that, in a closed circuit, the sum of the e.m.f.s is equal to the sum of the products of the current and resistance in each part of the circuit are of great importance in electrical engineering. Kirchhoff also made important contributions in the field of elasticity and spectrum analysis.

Gustav Robert Kirchhoff, the son of a lawyer, was born at Königsberg, Prussia, now part of the Soviet Union, on 12 March 1824. He left high school and entered Königsberg University where he studied theoretical physics under Neumann, who soon recognised Kirchhoff's talents and encouraged him to publish several papers. After getting his doctorate in 1848 he taught at Berlin University and then at Breslau where he met Bunsen. Both men moved to Heidelberg where they met Helmholtz with whom they formed such a successful team that students flocked from all over Germany and abroad to attend their lectures and laboratories. At about this time Kirchhoff became interested in elasticity and carried out investigations into the deflections and vibrations of flat plates and the 'large' deflections of thin bars.

Kirchhoff was also interested in the field of spectrum analysis in which he interpreted Fraunhoffer Lines and, with Bunsen, discovered the metals caesium and rubidium. Incidentally it was the almost non-luminous flame of Bunsen's famous burner which made such work possible. An interesting story concerns Kirchhoff's bank manager who, when told by Kirchhoff of the discovery of terrestrial metals on the sun, said, "Of what use is gold on the sun if I cannot get it down to earth?". Later, after Queen Victoria of England had presented Kirchhoff with a medal and a prize in gold sovereigns for work on the sun's spectrum, he took it to the bank manager and said, "Here is some gold from the sun!"

In 1868 Kirchhoff seriously injured his leg in an accident and this affected his general health. In 1875 he returned to Berlin University to the Chair of Physics, a post with less onerous duties. His health continued to deteriorate and he was forced to retire in 1884. He died in Berlin on 17 August 1887 at the age of 63.

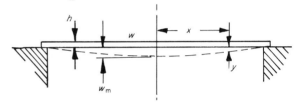

$$\text{Stress} \quad q = \frac{Eh^3}{12(1-\mu^2)} \left(\frac{\partial^4 w}{\partial x^4} + \frac{2 \partial^4 w}{\partial x^2 \partial y^2} + \frac{\partial^4 w}{\partial y^4} \right)$$

Kirchhoff's plate deflection equation

Joseph Louis Lagrange 1736-1813

Mathematician. Italian born, of French family

If his father had not lost most of his considerable fortune in high living and poor speculation, Lagrange would probably never have become interested in mathematics and his great contribution to theoretical mechanics would have been lost to the world.

Joseph Louis Lagrange was born on 25 January 1736 of a French family, in the northern Italian city of Turin. He showed an exceptional ability in mathematics at school, and by the age of 17 had studied and been deeply impressed by the memoirs of the English astronomer E. Halley, of comet fame, on the application of algebra to optics and astronomy. He was less impressed by the works of the Greek geometricians.

In 1755 at the age of 19, Lagrange was appointed Professor of Mathematics at the Artillery School of Turin where, with some of the most able students, he founded the Turin Academy of Science. In the same year he sent a paper containing his solution to the 'isoperimetric problem' to the renowned mathematician Leonhard Euler. This work, which dealt with geometric figures having the same perimeter, so impressed Euler that he withheld publication of his own paper on the subject considering his proof inferior to Lagrange's. Four years later Euler was instrumental in Lagrange's appointment as a member of the Berlin Academy of Science.

In 1766, the year of his marriage, at the invitation of

Frederick the Great and on the recommendation of Euler and D'Alembert, Lagrange succeeded Euler as professor at the Berlin Academy where he began to publish a long series of papers including his famous 'Mécanique analitique'. At the height of his fame his health deteriorated and he became subject to prolonged attacks of acute depression. Mathematicians and scientists in those days were, like musicians and painters, dependent upon patronage and, when Frederick died, conditions at the Berlin Academy deteriorated to such an extent that Lagrange decided to move to Paris. He was warmly received and given lodgings and facilities in the Louvre Palace and, more importantly, a generous grant by King Louis XVI. He was again the victim of severe melancholy and became disenchanted with mathematics turning instead to metaphysics, botany and chemistry. At about that time he remarried, his first wife having died some years earlier.

A few years later the Revolution came and with it the guillotining of many intellectuals including Lagrange's close friend Lavoisier. He decided that his life was in danger and made plans to leave the country but to his surprise, the new government, which had closed all the universities, offered him the professorship of the newly opened École Normale. He accepted as there was little alternative and he began to promote the proper training of teachers in France. He also set up the mathematics department at the new École Polytechnique and, under the new regime, perfected the metric system.

All this activity restored Lagrange's interest in mathematics and he devoted himself to writing and lecturing, his lectures being so inspired that they attracted teachers and professors as well as students. He wrote two books at this time *Fonctions analitique* and *Traité de la Résolution des équations numérique*. The last years of his life were occupied in a revision of his *Méchanique*, but before it's completion he died at the age of 77 in Paris.

Lagrange's masterpiece was undoubtedly his *Méchanique*, published a little over a century after that other great work in mechanics, Newton's *Principia*. It contains what are known as 'Lagrange's Equations' which are so useful in the study of applied mechanics. He was extremely proud of the fact that no diagrams, constructions or mechanical reasonings are included, only algebraic operations. The famous Irish Mathematician Hamilton referred to it as 'a scientific poem' but Lagrange had great difficulty in finding a publisher willing to handle such an abstruse work. He also contributed to many fields of mathematics and astronomy including work on the theory of equations, group theory and a study of partial differential equations. He interested himself in Elasticity and the Strength of Materials producing a study on the strength and deflection of struts 'Sur la figure des colonnes', in which he extended Euler's theories.

By all accounts Joseph Louis Lagrange was a modest man who did not seek public acclaim. The only portraits of him were sketches produced without his knowledge. He was undoubtedly one of the really great mathematicians of his time and one of that famous body of men including Monge, Prony, Fourier, Poisson and Navier who were associated with the École Polytechnique. His work on mechanics and to a lesser extent on elasticity has been of great value in the solution of many engineering problems.

Lagranges' dynamical equations:
$$\frac{d}{dt}\left(\frac{\partial T}{\partial q_i}\right) - \frac{\partial T}{\partial q_i} = Q_i, \ i = 1, 2, 3, \text{etc.}$$

Gabriel Lamé 1795-1870

French mathematician and engineer

Although Gauss considered Gabriel Lamé to be the foremost French mathematician of his day, others including Bertrand, praised him highly as an engineer. In general, French mathematicians thought him far too practically minded while scientists complained that he was too theoretical. But Lamé was brilliant at both; he thought that his mathematical work was of value to engineering and considered his most important contribution to be his development of curvilinear co-ordinates which he applied to many practical problems.

The mechanical engineer will come across the name of Lamé in connection with the well-known 'Lamé Equations' for determining the strength of 'thick' cylinders. The equations, for cylinders subject to internal pressure and for compound cylinders, were derived by Lamé as a result of practical experience when designing gun barrels for the Russian Government.

Gabriel Lamé was born in Tours, France on 22 July 1795. After showing an outstanding ability at mathematics at school he entered the École Polytechnique in Paris in 1813 and graduated in 1817; he then spent a further three years at the École des Mines. When he graduated his interest lay mainly in geometry, but before he had a chance to pursue this interest, he and his close friend and classmate Bénoit Clapeyron, were recommended to the Russian Government as two promising young engineers eminently suited to teach in the newly opened School of Highways and Transportation in St Petersburg (now Leningrad). There they taught physics, chemistry, mathematics and mechanics; they were also expected to design roads, bridges and tunnels in and around the city. The two young men even managed to find the time to publish works on mathematics and engineering. Several of the bridges they helped to design were the first suspension bridges to be constructed in Europe. To test the quality of Russian produced iron to be used in the construction of bridges, Lamé designed a special tensile testing machine.

In 1832, Lamé and Clapeyron returned to Paris following a deterioration in the relationships between France and Russia. With two other partners they founded an engineering firm, but after a few months Lamé resigned and accepted the Chair in Physics at the École Polytechnique where he stayed until 1844. He still managed to undertake a considerable amount of consulting work in engineering and was appointed Chief of the Mines. This exceedingly busy man also found time to plan and build two railways in Paris. In 1843 Lamé was elected a member of the Académie des Science and from 1844 to 1850 he was the examiner in mathematics for the University of Paris; he then became Professor of Mathematical Physics and Probability at the University, staying there until his retirement. His many books were widely read and valued by the students, but apparently he was by no means an outstanding lecturer.

Among the numerous papers which Lamé published, there were many on mathematics including those on curvilinear co-ordinates and differential geometry. There were also practical works on the design of bridges, artillery, gears, mines, tunnels and arches. He wrote a great deal on elasticity and was the first to advance the Theory of Failure based upon an Ultimate Tensile Stress.

In 1862 Lamé unfortunately lost his hearing and had to retire from the University and from public commitments, his general health deteriorated and he died on 1 May 1870 in Paris, six years after the death of his friend Clapeyron.

Lamé was yet another example, only too rare today, of a brilliant theoretician who was not averse to applying his theory to the practical needs of the day, nor did he tend to specialise in one field of technology as is the present custom.

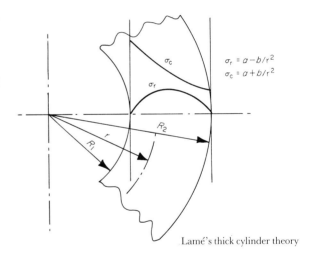

Lamé's thick cylinder theory

Pierre Simon Laplace 1749-1827

French mathematician

It is generally agreed that the mathematical and scientific work of Pierre Simon Laplace is second only to that of Isaac Newton. The greater part of his work was in the field of celestial mechanics where he showed himself a worthy successor to Newton. He was given his first big chance by D'Alembert, and was inspired by another first-class mathematician Lagrange, who, in a friendly way, was to be his greatest rival.

The name of Laplace occurs in both mechanical and electrical engineering; in Applied Mechanics for example, much use is made of Laplace Transforms, and in the study of electric and magnetic fields, the Laplace Equation is applied extensively.

Pierre Simon Laplace was born at Beaumont in Normandy, France on 22 March 1749 of impoverished farming stock who were unable to afford anything but the

most basic education for their obviously intelligent son. He showed such scholarship however that the generous neighbours paid to send him to a good school in Caen. He did well there and eventually obtained a teaching post at the Beaumont Military Academy. By the age of eighteen Laplace had completely mastered the subject of applied mathematics to date and had managed to obtain a letter of introduction to the great D'Alembert. He travelled to Paris to see the eminent mathematician but unfortunately failed to get an interview with him. The disappointed young man returned home and instead wrote a letter to D'Alembert in which he enclosed some of his own work on applied mechanics. The great man was so impressed that he replied immediately, offering Laplace a post at the Military School in Paris.

Laplace was to spend the rest of his life in Paris occupying a series of official posts and turning out a vast quantity of scientific papers and books on mathematics and astronomy.

He managed to survive the turbulent years of the French Revolution, the Napoleonic era, and the return of the Bourbons. He was honoured by Napoleon Bonaparte who once discussed astronomy with him on the battlefield; the returning Bourbons made him a Marquis.

The works of Laplace included a considerable amount on astronomy, and he published five great volumes on *'Mecanique Céleste'* which contained all the discoveries in astronomy since the time of Newton. His *Exposition du Système* was a beautifully written popular book on the Universe not containing any mathematics. The foundation for all his subsequent work on Probability Theory is to be found in his *Theorie Analytique des Probabilities*.

Laplace won many honours in his lifetime, Napoleon made him a member of the Senate, a Grand Officer of the Legion of Honour, and a Count of the Empire. He died in Paris on 5 March 1827 leaving an enormous amount of work which was to be of such great importance to the 19th century mathematicians. On his deathbed he echoed Sir Isaac Newton who compared his inadequacy to the child who finds the pretty pebble on the seashore, saying "What we know is minute, what we are ignorant of is vast."

Laplace transform of $f(t)$ is

$$F(p) = \int_{t=0}^{t=\infty} e^{-pt} f(t) \, dt$$

Jean Joseph Étienne Lenoir 1822-1900

French inventor

The story of Lenoir is one of a man of great ingenuity who, although he was recognised in his own time, lacked sufficient enterprise and business acumen to achieve the fame and monetary reward which he deserved.

Jean Joseph Étienne Lenoir, born on 12 January 1822 in Mussy-la-Ville, Luxemburg, taught himself engineering and chemistry and soon showed himself to be an outstanding inventor in many fields, including electric machines, railway signalling systems, the telegraphic transmission of pictures and internal combustion engines. He moved to France when he was sixteen and eventually became a French citizen.

For over half a century men like Savery, Newcomen, Watt and Trevethick had developed the steam engine, with its *external* source of heat, where a reasonable degree of efficiency and reliability had been obtained. For a much longer period, over two centuries in fact, beginning with Christian Huygens and Denis Papin, men had striven to obtain power from an explosion *inside* the cylinder of an engine, and Lenoir was to be the first to achieve success. He took out a patent for his engine in 1860. Since highly volatile liquid fuels were not available in those days, Lenoir used illuminating gas, and in the next five years he sold over 300 of his engines in sizes ranging from $\frac{1}{3}$ to 3 hp. These double-acting engines were not very efficient and suffered from low compression, excessive heat loss, and incomplete expansion of the hot gases. However, in 1860 Lenoir fitted one of them into a carriage and thus produced the first motor car. He later made a reasonably good four-cylinder marine engine and incidentally was the first man to power a boat with an internal combustion engine.

One day in 1860 a young German clerk, Niklaus Otto, read a newspaper article on Lenoir's new engine and as a result became intensely interested in the internal combustion engine. In 1864, with the help of the enterprising industrialist Eugen Langen, the firm of Niklaus Otto and Company was founded to market a much improved engine designed by Otto. Poor Lenoir could not possibly compete with the ingenuity of Otto backed by the business efficiency of Langen and had to give up his firm. He died poor and unhonoured in the country of his adoption, at Varenne-St-Hilaire, on 14 August 1900. Since his death Lenoir's reputation has improved enormously and he is now recognised as the true inventor of the first internal combustion engine although the principle upon which such engines work is known as the Otto Cycle.

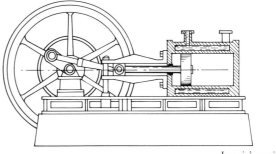
Lenoir's engine

Leonardo da Vinci 1452-1519

Florentine engineer, anatomist, painter, musician

Leonardo da Vinci is not mentioned often in engineering textbooks, nor is he responsible for any theory or law of value to the engineer. It would be unusual however if he were not referred to at one time or another by lecturers in an engineering course. This brilliant Florentine stands out at perhaps the greatest genius of the Renaissance. As well as dominating the world of painting, sculpture and architecture Leonardo da Vinci was a highly talented inventor, scientist and engineer. He was responsible for the invention of a vast number of ingenious devices, many of which could not be developed in his day because of the lack of suitable materials and manufacturing techniques. Engineers of all fields, young and old, cannot fail to be fascinated and indeed inspired by his technological achievements.

Leonardo had a rather unfortunate start in life; he was the illegitimate son of a young member of the nobility, Piero da Vinci; his mother was a pretty country girl called Catherine, from a neighbouring village. He was born in the small town of Vinci not far from Florence on 15 April 1452, where later both his parents married members of their own class, but Leonardo's father agreed to bring him up with all the advantages of a legitimate son of a respected member of the ruling Council of Florence.

As a child, Leonardo showed a great aptitude for drawing so, at the age of 14, he was sent to study painting, music and verse under the famous Verrochio, he was also enrolled in the painter's guild at Florence. After studying the arts for 14 years, during which time his interest turned to music rather than painting or sculpture, Leonardo travelled around Italy for about four years to acquire greater experience in the arts. At the age of 32 he ceased his wanderings and accepted the post of Master of Revels at the court of the Prince of Milan. He now began to earn his living as a writer of popular songs and as a producer of the most extravagant musical shows. These were produced for weddings and other social occasions at court. Somehow Leonardo still managed to find time for drawing and painting and it was during this period that he painted that great masterpiece the 'Last Supper' and produced many of his wonderful anatomical drawings.

It was not until he was 54, an age when many now consider early retirement, that Leonardo da Vinci began his career as an engineer. His first commission was as a designer of the engineering works for the great Lombard Canal system. Cesare Borgia then hired him as a military engineer to design fortresses, rapidly-assembled bridges, guns, tunnelling equipment, etc. About that time he designed a number of instruments including a wind velocity meter and a device for measuring the speed of ships.

In 1515 at the age of 63, Leonardo was invited to France by King Francis I who installed him in a chateau near Amboise on the Loire. He continued to carry out his engineering works acting as an adviser to the king until the day of his death on 2 May 1519.

Leonardo da Vinci was undoubtedly centuries ahead of his time in many of his technological inventions. His suggestions for man-powered flight and the invention of the parachute were not to be realised for five centuries, and he anticipated the hot-air balloon when as a party trick he filled hollow wax figures with hot air and let them float around a room. He wrote, "A bird is a machine working according to mathematical laws, which it is within the capacity of man to reproduce. Such a machine is lacking in nothing but the life of the bird and this life must needs be supplied by man.'' Apropos of the parachute he wrote, ''If a man has a tent made of linen, twelve braccia (25 ft) across and twelve braccia deep, he will be able to throw himself down from a height without sustaining injury.''

The sketches of Leonardo show mechanisms in great detail with cranks, connecting-rods, gears, cams, and universal joints; there are pumps, dredgers, pile-drivers, and windmills. There are tools for screw and file cutting, also textile machinery. He anticipated the centrifugal pump and the hydraulic press.

The work of previous investigators was ignored invariably by Leonardo, who believed in accepting only that which he himself had observed. His advice to scientists and engineers is well worth noting. ''Before you make a result a general rule, test it by experiment two or three times and see if you get the same result each time.'' The expression 'Tested by experiment' appears many times in his writings.

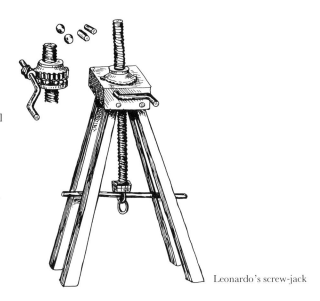

Leonardo's screw-jack

Ernst Mach 1836-1916

Austrian physicist and philosopher

Ever since the 'sound barrier' was broken in flight, the man-in-the-street has been familiar with the words 'supersonic' and 'Mach Number'. Very few will realise that there actually was a man called Mach or even know exactly what the Mach Number is.

Ernst Mach was born in Turas in Moravia, now part of Czechoslavakia, on 18 February 1836, one hundred and ten years before man flew at 'Mach One'. His father, a schoolmaster, moved to Vienna when Ernst was a baby, and it was there that he was educated, first at school and then at the University. He graduated and was appointed as Professor of Mathematics at Graz at only 26 years of age. After three years he accepted the Chair of Physics at Prague where he stayed until he was 57; it was in Prague that he carried out his studies on aerodynamics and supersonics.

In the mind of the engineer Ernst Mach is associated mainly with the flow of fluids at high velocity and Mach Number, but he is also well known for his interest in the philosophy of science. He was very much concerned with the abolition of metaphysics from science and believed that all assumptions not controlled by actual experience were valueless. As an example of this assertion he gave the concept of absolute 'space and time' of Newtonian physics. He completely rejected atomism and thought of the atom merely as a model and not to be confused with ordinary matter.

Mach favoured the doctrine of 'sensationalism' which holds that all knowledge of the world comes from one's sensations such as colours, spaces, times, etc. and not from 'things'. His views in what was called the 'Vienna Circle' were important to 'Positivists' and greatly influenced many physicists of the day, leading for example to quantum mechanics. His famous assertion that space did not exist but only the matter in it, was called 'Mach's Principle' by Albert Einstein.

In his experimental work Mach was interested particularly in the flow of high velocity gases over objects and inside pipes and ducts. He was intrigued by the sudden change in the pattern of flow as the speed approached that of sound in a gas. Between 1873 and 1893, working with collaborators including his son Ludwig, he devised and perfected optical and photographic techniques for the study of sound waves produced by high velocity flow, the flight of projectiles, meteorites, explosions and gas jets. In 1887 he published a famous paper on 'Supersonics' based upon an investigation carried out for and sponsored by the Royal Austrian Navy.

The quantity now known as Mach Number, is one of many such numbers referred to as 'dimensionless' and is of great interest to engineers. It is the ratio of the velocity of an object moving in a gas, or in the case of a stationary object the velocity of the gas itself, to the velocity of sound in the gas. If the velocity is twice that of sound for example we say that the flow is at 'Mach 2', or that $Ma = 2$. When the velocity is greater than Mach 1, that is supersonic, a pressure wave called the Shock Wave or Mach Wave is produced in front of the object, analagous to the bow wave of a ship. The angle of this wave is known as the Mach Angle. Curiously enough it was not until 1929 that Ackeret suggested the use of the term Mach Number.

In 1893 Mach went to the University of Vienna as Professor of Physics, where he stayed until his retirement in 1901 after suffering a stroke, when he was succeeded by the noted physicist Boltzmann. Mach served in the Austrian House of Lords in his latter years and died at Haar near Munich on 19 February 1916.

Ernst Mach the great physicist and philosopher was by no means an engineer, yet he laid a foundation for high speed flight, the immense value of which has only been fully realised since the design of supersonic aircraft gained such impetus in recent years.

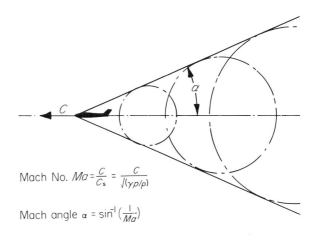

Mach No. $Ma = \dfrac{c}{c_s} = \dfrac{c}{\sqrt{(\gamma p/\rho)}}$

Mach angle $\alpha = \sin^{-1}\left(\dfrac{1}{Ma}\right)$

Heinrich Gustav Magnus 1802-1870

German chemist and physicist

The 'Magnus Effect' is a phenomenon well known to students of mechanical engineering, which has been brought to the public eye by sports commentators who refer to a football being 'bent' in flight or to the 'swing' of a cricket ball. When a football, cricket ball, or golf ball is given spin it will tend to move through the air in a curved path. If the spin is about a horizontal axis the ball will curve upwards for 'bottom spin' and downwards for 'top spin'. Spin about a vertical axis will make the ball curve to left or right, so useful of course to a bowler in cricket.

The same effect is obtained with a rotating cylinder moving through the air with its axis at right angles to the airflow. A lift force is produced similar to that for a wing. Non-circular shapes exhibit the same lift when rotated, toy kites with rotating aerofoils have been made which work on this principle.

Heinrich Gustav Magnus who first investigated this phenomenon was born in Berlin on 2 May 1802, the son of a prosperous owner of a large trading company. Heinrich was given the very best of private education and was then sent to the University in Berlin in 1822. There he specialised in chemistry, his first paper being on the metals iron, cobalt and nickel. He obtained his Doctorate for a dissertation on tellurium in 1827 and then moved to Sweden to work under the famous Swedish chemist Berzelius who became his lifelong friend and adviser. Magnus returned to Berlin in 1828 where he met and married Bertha Humblot in 1840. He became Professor of Technology and Physics in the University of Berlin in 1845 and served for two years as Rector, during which time he founded the German Chemical Society.

The researches carried out by Magnus were many and varied but were concentrated mainly in the field of chemistry, such as mineral analysis, inorganic chemistry, polymerisation, the oxygen and carbon dioxide content of

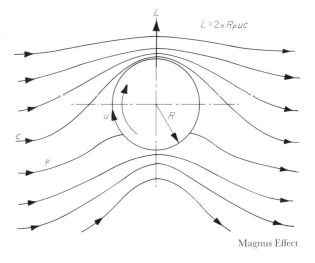

Magnus Effect

blood, agricultural chemistry and so on. He was interested particularly in the manufacture of sulphuric acid.

After this tremendous interest in chemistry and for no apparent reason, Magnus seemed to turn entirely to physics where again he worked on a vast range of topics, such as thermal expansion of gases, the boiling of liquids, vapour formation, electrolysis (Magnus' Rule), thermoelectricity, optics, magnetism, mechanics, and of course fluid mechanics and aerodynamics. It was in connection with this last topic that he studied the flow of air over rotating cylinders and in 1853 discovered the mathematical theory from which the lift force could be calculated, known as the Magnus Effect.

An interesting application of Magnus Effect, often mentioned in textbooks on Fluid Mechanics, was in the famous Flettner Rotor Ship, a converted schooner using two huge vertical rotating cylinders for propulsion, instead of sails. The cylinders were about 3 metres in diameter and 20 metres in height and could be made to rotate in either direction by means of a small engine. The ship was propelled by the force produced by the wind and the rotation. This meant of course that the wind was on the beam. This rotor ship, named the 'Buchau', actually crossed the North Sea and returned successfully. In 1926 the same ship, with three towers and renamed the 'Baden-Baden' crossed the Atlantic successfully. No further attempts were made using this method of propulsion, although the Magnus Effect was applied also to windmill sails by Flettner.

Magnus stayed at Berlin University until his retirement, and he died there on 4 April 1870. Although he was by no means a great original thinker, Magnus carried out researches which yielded much information of value to the engineer and physicist.

James Clerk Maxwell 1831-1879

Scottish mathematician and physicist

James Clerk Maxwell is universally accepted as the greatest theoretical physicist of the nineteenth century; he possessed a wonderful mathematical ability and a complete grasp of physical reality together with a brilliant imagination. He was able to make great advances in science without recourse to preconceived ideas about the workings of nature.

Maxwell's main task was the mathematical interpretation of Faraday's concept of the electromagnetic field which had been deduced by laboratory experiments. In carrying out this task to its conclusion he gave the world the electromagnetic theory of light which paved the way for the discoveries of Rudolf Hertz 23 years later which led to the invention of radio and the theory of relativity. Maxwell also made significant contributions to the field of elasticity and, after being made a gift of two prisms by Nichols was the first to apply methods of photo-elastic stress analysis which are familiar to the engineering student today. He developed 'frozen-stress' techniques, important for three-dimensional stress analysis and in the field of structures, and in 1864, he derived a method for determining loads in roof trusses which, because of the high mathematical content, was rarely used by engineers. Ten years later the method was rediscovered by Mohr and simplified to give what is referred to as the Maxwell-Mohr Method. Maxwell is of course renowned for his memoir on the kinetic theory of gases and statistical mechanics, both of which led to the quantum theory.

James Clerk Maxwell was born in Edinburgh on 13 June 1831 but spent most of his childhood at his parent's country house 'Glenlair' on their estate at Middlebie. His mother died from cancer when James was only eight years old and, like the other great Scottish physicist Lord Kelvin, was brought up by his father. At the age of ten, James was sent to the Edinburgh Academy where his broad Galloway accent and the rather fancy clothes in which his father had unfortunately dressed him, prompted the other boys to bully him. They called him 'Dafty' and beat him so that he arrived home at the end of the first day bleeding and in rags.

Maxwell showed no outstanding ability until he was introduced to a course in mathematics at the age of thirteen, he then made such swift progress that he won a gold medal for mathematics at the age of fourteen. At sixteen he entered Edinburgh University where he made such an impression that he was allowed the run of the chemical and physics laboratories and was able to carry out any research in which he was interested. He entered Trinity College, Cambridge in 1850 and was appointed Professor of Natural Philosophy at Aberdeen in 1856. In 1860 he was appointed Professor of Natural History and Astronomy at King's College, London.

While at King's College, Maxwell produced his important memoir on the kinetic theory of gases and statistical mechanics. It was in connection with this work that he postulated his famous 'Maxwell's Demon.' This was an imaginary creature who, by opening and shutting a tiny door between two enclosed volumes of gas, could theoretically admit slower (colder) molecules to one volume and faster (hotter) molecules to the other volume thus reversing the tendency for increased disorder (entropy) and so violating the Second Law of Thermodynamics. During this time he produced his work on elasticity and electromagnetism.

The name of this great scientist is linked with many theories and inventions. In physics there are the Maxwell-Boltzmann distribution law, Maxwell's Relationship, and Maxwell's Thermodynamic Relations. In electricity, the cgs unit of magnetic flux (the maxwell), the Maxwell Bridge for inductance and capacitance, Maxwell's Rule, Maxwell's circulating current, and Maxwell's circuital theorem. In optics, the Maxwellian viewing system, Maxwell primaries (the colours red, green and blue-violet) and Maxwell's experiment on colours. In engineering there is Maxwell's theorem for elastic structures and in mathematics Maxwell's field equations.

On the death of his father in 1865, Maxwell retired from King's College to return to his estates in Scotland where he devoted himself to research and writing his great treatise on electro-magnetism. In 1871 he was persuaded to accept the first Cavendish Professorship at Cambridge where he designed the new Cavendish Laboratories.

This great scientist who had produced vast quantities of work in so many fields, led the way to all the inventions which were to come in the field of electromagnetic radiation. He continued to work with unabated enthusiasm until the time of his death, which occurred after some stomach trouble, ignored because of pressure of work. He suddenly became seriously ill and died at Cambridge on 5 November 1879 at the age of 48.

Maxwell's field equations:

$$\text{div } B = 0; \quad \text{curl } H = \frac{\partial D}{\partial t} + j;$$

$$\text{div } D = \rho; \quad \text{curl } E = -\frac{\partial B}{\partial t}.$$

Christian Otto Mohr 1835-1918

German civil engineer

In Strengths of Materials, problems on Complex Stress have a tendency to fill the average engineering student with more than a little apprehension. The difficulties may be eased considerably when the student applies Mohr's Circle of Stress. A Strain Circle and a circle for determining principal Second Moments of Area were also produced by Mohr together with a theorem for statically indeterminate beams and continuous beams.

Christian Otto Mohr was born in Wesselburen, a village on the bleak coast of the Heligoland Bight in Schleswig-Holstein, on 8 October 1835. Descended from a long line of Holstein landowners, he commenced his studies in engineering at the Polytechnic Institute at Hanover in 1851 and after qualifying, worked as a civil engineer first in Hanover and then in Oldenburg in Schleswig-Holstein. He was apparently an exceedingly good bridge builder but in 1867 he decided to give up practising as an engineer and took up an appointment as Professor of Mechanical and Civil Engineering, at the Technische Hochschule in Stuttgart. Six years later he moved to Dresden as Professor of Engineering and remained there until his retirement.

During his teaching career, Mohr carried out a tremendous amount of research into the theory of structures and elasticity and continued to do so until long after his retirement in his eighties. Unlike many great men of science, Mohr was a remarkably good lecturer and his lectures were, like his research papers, of amazing clarity and brevity. He was impressively tall and good-looking with a proud and haughty manner no doubt inherited from his Danish landowner forebears.

Although Mohr produced a large number of scientific papers, he only wrote one textbook, *Technische Mechanik*, which is discussed with his other work in Stephen Timoshenko's *History of Strengths of Materials*. His work on stresses in indeterminate frames was published in 1874 while work relating to his Stress Circle was completed in 1882. The

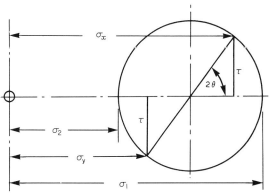

Mohr's Circle of Stress

theory of failure based upon the Stress Circle is widely used by engineers. The work on continuous beams follows Clapeyron, but it is a distinct improvement and includes the effect of variation in height of the supports. His use of what are known as 'Influence Lines' in continuous beam problems greatly eases the problems in the determination of the deflections in beams of varying second moment of area.

Mohr taught at the University of Dresden until his retirement at the age of 65, and spent his remaining years quietly pursuing his researches until his death in Dresden on 2 October 1918.

Richard Mollier 1863-1935

Italian-born German thermodynamicist

The introduction of the Enthalpy-Entropy Diagram towards the end of the 19th century proved to be a great boon to the steam turbine designer. It enabled him to see exactly what should happen in each stage of the turbine and thus permit rapid calculation of the output performance and cycle efficiency. This diagram, also known as the Mollier Chart, after the German engineer whose main interest was the production of thermodynamic diagrams and charts is extremely useful to the designer of heat engines.

Richard Mollier was the son of Edouard Mollier, a Rhinelander who was a naval engineer and later the director of a large machine company in Trieste in Italy. It was there that Richard was born on 30 November 1863 and was sent to the German Gymnasium where he graduated with honours in 1882. He next attended the University of Graz, then the University of Munich, where he studied mathematics and physics. Finally he went to the Technische Hochschule at Munich where he studied under the famous Carl von Linde, graduating in 1888. After two years' engineering experience in his father's factory in Trieste, Mollier became assistant to Professor Schrotes at the Munich Technische Hochschule.

Mollier worked on thermodynamics applied to machines and in 1892 produced several thermodynamic diagrams, including the famous Mollier Chart, together with an air-vapour mixture diagram. He obtained his Doctorate for work on the entropy of vapours in 1895 and, following a brief stay at Göttingen to introduce Technical Physics to the curriculum, he was appointed Professor of Theory of Machines and Director of the Machines Laboratory at Dresden Technische Hochschule in 1897. He was to spend the remainder of his life there.

Mollier was a kindly, retiring man who took his lecturing extremely seriously; he would not use notes but would memorise the work and then deliver it with great clarity and simplicity. In this way he achieved a spontaneity which he hoped would create greater interest among the students. Among his students was the illustrious Nusselt, the expert on heat transfer, and his own sister Hilde.

Richard Mollier achieved world-wide fame for his work on thermodynamics which was to prove of such lasting value to the practical engineer. He died at Dresden on 13 March 1935.

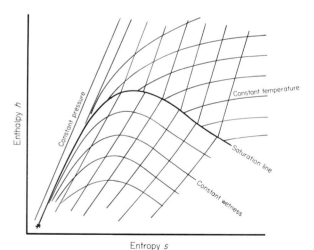

Mollier h-s chart

Sir Isaac Newton 1642-1727

English scientist
and mathematician

Possibly the greatest scientist of all time, Isaac Newton started life most inauspiciously in the same year that Galileo died. He was born prematurely in the most humble circumstances in the village of Woolsthorpe in Lincolnshire and for some time after his birth it was doubtful whether he would live. His birthday was Christmas Day 1642 according to the old Julian Calendar then in use, or 4 January 1643 in the present Gregorian Calendar.

Shortly after Newton's birth his widowed mother remarried and he was left in the care of grandparents. At the age of twelve he was sent to school at Grantham where he seemed to show little sign of outstanding academic ability but spent a considerable amount of time making kites, waterclocks and sundials; he certainly did not seem to have been well taught. After leaving school he spent some time on his mother's farm until in 1661 he was fortunate enough to be accepted for Cambridge University through the influence of an uncle.

After graduating without distinction in 1665, Newton decided to stay on at Cambridge but was sent home because of the Plague which was then ravaging London. At home he began to study the works of all the great mathematicians and became especially interested in optics and applied mathematics. It was during this period that the famous 'apple' incident occured; if anyone doubts the truth of the story they are recommended to refer to Newton's official biography where it is recorded. At this time Newton began to develop the Calculus and the Binomial Theorem and to speculate on the force which kept the moon in its orbit, and the inverse square law.

Although the spectrum was known to the ancient Greeks, Newton was the first to observe that the disc of light from the sun could be made to produce a 'rectangle' of coloured bands, showing conclusively that the colours were physically separated. The real stroke of genius was to use a second prism to restore the white disc of light.

In 1667, at the age of 28, Newton returned to Cambridge as Lucasian Professor of Mathematics, a post in which he was only supposed to deliver eight lectures in a year. It is recorded that he gave these lectures rather poorly and spent the rest of his time engaged in research on a great variety of topics. In 1672 he became a member of the Royal Society where he met many eminent scientists of the day including Robert Hooke who was the first Curator of the newly formed Society. The two men did not take to each other and had violent disagreements over Newton's experiments on light. This unfortunately led to a life-long enmity between the scientists. It may have been the result of a childhood without proper parents that made Newton into an unloved and lonely figure, for he never married and became increasingly suspicious of other scientists whom he thought were trying to steal his work.

Isaac Newton's great work, *Principia Mathematica*, which he wrote in only 18 months, was published in 1687. It embodied all his works on mechanics and is considered by many to be the greatest scientific work ever produced. It contains a number of rules under the heading 'Reasoning in Philosophy'. Rule 1 says, "We are to admit no more causes of natural things than such as are both true and sufficient to explain their appearances."

In his late forties Newton's mind seemed to be deteriorating. He indulged in a popular but fruitless pastime of those days, trying to calculate the date of the Creation, which he found to be 3500 B.C., and spent a great deal of time and money trying to find ways of transmuting base metal to gold. There is some evidence that experiments with mercury led to mercury poisoning which can seriously affect the mental faculties.

At the age of 49 Newton suffered a complete nervous breakdown and retired to the country for about two years during which time he must have at least partially recovered his mental faculties, for sometime later Leibnitz the co-discoverer of Calculus, sent him a difficult mathematical problem to solve. This attempt to stump the ailing mathematician failed when Newton solved the problem in an afternoon.

Wellwishers managed to get Newton an honorary post as Warden of the Royal Mint in 1696 and he was later made Master of the Mint in 1699. These well-meaning but misguided friends however only kept the now partially

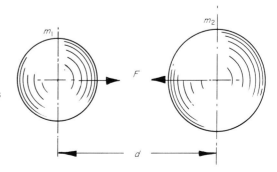

$F = \dfrac{Gm_1 m_2}{d^2}$

F = Force of attraction
G = Gravitational constant
m_1, m_2 = Masses
d = Distance between centres of masses

recovered Newton away from his work, with who knows what loss to science. It did mean that he was able to leave the considerable sum of £30,000 when he died.

In 1703 Newton was elected President of the Royal Society and in 1705 he was knighted by Queen Anne, the first scientist ever to be so honoured. He died at the age of 85 on 20 March 1727 and was buried with great ceremony in Westminster Abbey. Shortly before his death Newton had said, "I seem to have been like a boy playing on the seashore . . . and finding a smoother pebble or a prettier shell than ordinary, whilst the great ocean of truth lay all undiscovered before me."

In a course on Mechanical Engineering there is hardly a single lecture on engineering topics in which the work of Isaac Newton is not touched upon. The definitions of Force, Mass, and Momentum, and his Three Laws of Motion crop up continually. In heat transfer there is Newton's Law of Cooling and in Fluid Mechanics Newton's definition of Viscosity and reference to non-Newtonian fluids. It is fitting that the unit of force in the Système International should be the 'newton' (symbol N) which incidentally happens to be about the 'weight' of an average apple.

Ernst Kraft Wilhelm Nusselt 1882-1957

German engineer

The tremendous increase in the use of high-powered machinery in the early 20th century created immense problems in the calculation of the heating and cooling of liquids and gases. Unfortunately the only methods available for the solution of heat-transfer problems were those due to Fourier, which applied only to heat flow in solids and were useless for fluids where heat transfer existed by convection alone.

The first contribution of any significance to the analytical treatment of heat transfer by convection was made by the German engineer Ernst Nusselt. Instead of waiting for a complete mathematical solution he applied the principles of Dimensional Analysis, so well known to engineers today, where he showed in an amazingly simple way the form of the solutions; thus it was possible to use the very limited data available in those days.

Ernst Nusselt was born in the German city of Nuremberg on 25 November 1882, the son of a factory owner Johannes and Pauline (Fuchs) Nusselt. His early education took place in Nuremberg until he entered the Technische Hochschule in Munich and studied Mechanical Engineering. He then moved to the Charlottenberg in Berlin but returned to Munich to obtain a Mechanical Engineering Diploma. There he worked as an assistant to Knoblauch who had taught the expert on heat transfer Ernst Schmidt whose name is given to the Schmidt Number. Nusselt was awarded his Doctorate in 1907 and then worked alternatively in teaching and industry, during one period with the Swiss engine firm of Sulzer. He finally returned to Munich Technische Hochschule where he had been appointed to the Chair in Mechanics.

Of the large number of papers published by Nusselt, two are of particular importance. The first is 'Basic Laws of Heat Transfer', in which the large number of variables found in the study of heat flow in fluids are reduced to non-dimensional groups named after their investigators, such as Reynold's Number, Prandtl Number, Grashof Number and of course Nusselt Number. The second paper has the title 'Film Condensation of Steam' and is of great value to present-day engineers. In 1930 Nusselt did some important work on the similarity of the transfer of heat to the transfer of mass.

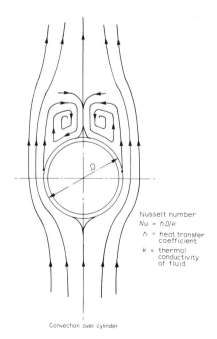

Nusselt number
$Nu = hD/k$
h = heat transfer coefficient
k = thermal conductivity of fluid

Convection over cylinder

What of the man himself? Nusselt had married Susanna Thurmer when he was 35 years old and teaching at Dresden, and they had two daughters and a son. Nusselt was by all accounts a very quiet, softly spoken, reserved man but he possessed tremendous energy. He was a keen mountain climber and attacked a mountain peak with the same enthusiasm and methodical manner as he did his scientific work. He was extremely cautious in his publications however and did not make the same impact as other German scientists of the 1930s and 1940s. He lacked charisma and in common with many other brilliant scientists was a rather poor lecturer.

The fact that he suffered for many years from a chronic internal complaint did not help matters. In 1947 his only son Dietrich fell to his death while mountaineering and the shocked Nusselt retired from his post and did little more work. He lived in seclusion in Munich until his death on 1 September 1957.

The name of Nusselt will be forever remembered by countless numbers of engineering students'whose studies include Heat Transfer. There they will find it linked with those two other great names, Reynolds and Prandtl in the form of the non-dimensional quantities, *Nu, Re,* and *Pr.*

Georg Simon Ohm 1789-1854

German physicist

George Simon Ohm is known for his discovery of the famous law which bears his name, also in connection with the unit of electrical resistance, the ohm. Although Ohm's Law seems very simple and straightforward nowadays and can easily be demonstrated with quite basic equipment, the publication of Ohm's paper, 'Die Galvanischette' in 1827 was truly epoch-making.

Ohm owed his skill at constructing his superb apparatus to tuition received from his locksmith father from Erlangen in Bavaria, where he was born on 16 March 1789. He went to school there and later taught mathematics and physics at several schools in the neighbourhood before returning to Erlangen for more advanced studies in science at the University. In 1817 at the age of 28 he was appointed Head of Department in Physics at the Polytechnique Institute of Cologne where he began to study the flow of electricity through various materials. It was there that he carried out his famous experiments, entirely on his own, using apparatus which he had built himself.

In his experiments, Ohm found that the voltage from the voltaic cell he was using varied considerably as the current was varied and he was obliged to use a thermocouple to measure the current. He discovered the empirical law relating the voltage and current in a circuit which applies very closely for good conductors such as metals. He announced the law in 1826, with little response, then repeated his experiments in 1827, this time taking into account the effect of internal resistance of the battery which thus solved the problem of voltage variation. He also suggested an analogy between the flow of electricity in a conductor and the flow of a fluid in a pipe.

Ohm's work was given a very poor reception universally, and many scientists even called it a fantasy. He lived in obscurity and in poor financial circumstances, his work disregarded until, in 1841, it was recognised in Great Britain. He was awarded a medal by the Royal Society who also made him a foreign member the following year.

In 1843, using Fourier's method of analysis, Ohm published an important paper on the reception of sound in the human ear and the ear's ability to separate the individual frequencies. Again his work was completely ignored until his theories were rediscovered by Helmholtz.

Germany rewarded this great physicist rather belatedly, by his appointment to the Chair of Physics at Munich University only two years before his death in Munich on 7 July, 1854.

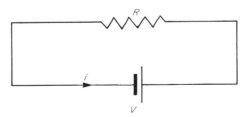

Ohm's Law $i = V/R$

Niklaus August Otto 1832-1891

German engineer

A shy, retiring clerk, son of a farmer and posthouse keeper, chanced to see a newspaper article about the wonderful new engine built by Étienne Lenoir. He was so interested that he decided to devote his life to producing engines which would supersede the old fashioned steam engine.

This young man was Niklaus August Otto, born in the small town of Holzhaus, Nassau in Germany on 14 June 1832. He left school at sixteen and was put into commerce by his parents in spite of his childhood interest in science and engineering. Thirteen years later Otto was spending all his spare time and money studying science and technology to such an extent that his friends nicknamed him Grübler– 'thinker'. By the end of 1861, Otto had at last managed to produce a small four-stroke internal combustion engine. He had recognised the value of compressing the mixture of air and fuel before ignition near top dead centre, and the use of separate expansion, exhaust and induction strokes. The engine unfortunately proved to be unsatisfactory because of the violence of the explosions which caused extremely rough running, so he turned his attention to the atmospheric gas engine of the Lenoir type.

In 1864, Otto was fortunate in meeting Eugen Langen, an industrialist, who helped him to form the firm of Niklaus August Otto and Company. The atmospheric engine was shown in the Paris Exhibition of 1867 where it won a gold medal in competition with fourteen French gas engines and showed an efficiency three times that of its nearest rival. After financial difficulties nearly closed the firm, more capital was raised and a new factory, the Gasmotorenfabrik, was built near Cologne in 1869. Otto attended to the business side of things and left the engineering side to two able recruits Maybach and Daimler. Fifteen years after his first attempt to produce a four-stroke engine, Otto produced another and patented the invention in 1877. In the words of one who saw the demonstration 8 hp engine "It ran so elegantly and beautifully that it would have given joy to an angel to watch it."

Otto's new engine was a tremendous success and others sought to find ways of 'cashing in' on the new idea. His patent, which covered both two- and four-stroke working, was contested and finally in 1886 the patent was invalidated after the discovery of papers describing the cycle by Beau de Rochas which had lain dormant for many years. It was in fact two years prior to Otto's meeting with, Langen, i.e. in 1862, that Alphonse Beau de Rochas took out a patent describing the main details of a four-stroke cycle but it did not include any practical suggestions as to how it might be produced. No one will ever know whether or not Otto knew of its existence. However over 3000 of the so-called 'silent Otto engines' were sold in the first ten years of manufacture and the firm had plenty of work on hand modifying the engine so that it ran successfully on liquid fuels such as petrol. Marine engines were built and designs for engines for locomotives and road vehicles were initiated.

Niklaus Otto died on 26 January 1891, modest and retiring to the end. Although due credit must be given to Lenoir for making the first working internal combustion engine, it was Otto, a truly great engineer, who developed it into the prototype for all modern internal combustion engines.

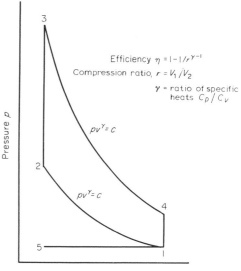

Efficiency $\eta = 1 - 1/r^{\gamma-1}$
Compression ratio, $r = V_1/V_2$
γ = ratio of specific heats C_p/C_v

Charles Algernon Parsons 1854-1931

British engineer

A primitive form of steam turbine was made by Hero of Alexandria nearly 2000 years ago, and in 1629 the Italian Giovanni Branca had produced at least a drawing of a steam-turbine-driven windmill which was probably not practicable. In 1815 Richard Trevethick experimented with a 15 ft diameter 'whirling engine' without much success. No one however managed to produce a workable steam turbine until the 1880s, by which time there had arisen a great demand for high power at high speeds, which was unobtainable from the reciprocating steam engine. The patent of Charles Parsons of 1884 relating to the steam turbine therefore heralded the advent of an era of undreamt of power produced at high speed.

Charles Algernon Parsons was born in London on 13 June 1854, the son of William Parsons, Third Earl of Rosse, distinguished astronomer and President of the Royal Society. Charles was brought up in wealth and comfort at the family seat at Birr Castle in Parsonstown, Ireland where he had the run of his father's extensive workshops and foundry. He received a private education at home as his father scorned education at schools, and was well tutored in both mathematics and physics. During the summer months the whole family put to sea in the Earl's luxurious yacht and the children's tutors accompanied them. Charles gained good experience in both seamanship and navigation, both of which were to prove of great value to him later in his career. At home, he made all kinds of mechanical devices, a 4 hp steam engine, an air gun, a sounding device for his father's yacht and even a small steam car. Lord Rosse had, of course, built his famous telescope called the Leviathan; this was immense, a 72-inch refractor of tremendous length, which is said to have cost over £30 000 in 1845 and through which galaxies could be observed for the first time. Young Charles took great pleasure in helping his father to operate what was at that time the biggest telescope in the world.

In 1872, Charles left these idyllic surroundings for Dublin University and after one year went to Cambridge where, because in those days there was no engineering course, he studied mathematics and physics. While he was at Cambridge, he designed and made a rotary steam engine; he was obviously already intrigued by the rotary principle. On leaving Cambridge, Parsons was sent to work as a student apprentice at the Elswick, Newcastle-upon-Tyne factory of Lord Armstrong where his father was sure he would gain good practical experience in engineering. There Parsons found time to carry out some of his own experiments, including work on rocket-propelled torpedoes.

After his apprenticeship, Parsons joined the Gateshead-on-Tyne firm of Clarke-Chapmans as a partner, where the firm at that time was extremely interested in the newly invented electric lighting of Joseph Swan of Sunderland, who was then working in Gateshead at a nearby works. Clarke-Chapmans were producing marine engines and were hoping to develop one to drive a dynamo, but Parsons, already interested in the problem, thought that a turbine was the best solution and took out a patent for a steam turbine of original design in 1884.

Parson's turbine used a large number of stages of alternate fixed and rotating blades through which the steam was gradually expanded, as opposed to the de Laval type of turbine in which the expansion was achieved in a single stage. This resulted in a lower speed of rotation and a rotor of much smaller diameter, but his first turbine produced a mere 10 hp at the very high speed of 18 000 rev/min. To utilise the energy of the exhaust steam, Parsons introduced a condenser operating under a high vacuum, and he also began to use high-pressure superheated steam in order to increase the efficiency. In 1888, the first ever turbine-driven generating set was installed in the Forth Banks Power Station in Newcastle-upon-Tyne with four 75 kW machines running at 4800 rev/min. Parsons was eager to build much larger turbines but Clarke-Chapmans were not interested so he opened his own factory at Heaton, Newcastle-upon-Tyne where for some years he had to make radial flow instead of axial flow turbines until his old firm of Clarke-Chapmans allowed him to use his own patent. By 1900 he was manufacturing turbines of over 1000 kW in his own factory.

Parsons became interested in applying the steam turbine to ship propulsion and in 1897 his firm completed the experimental 44-ton, steam-turbine-powered launch the 'Turbinia'. He used a series of model hulls from a few inches to 6 ft in length, some of them driven by twisted rubber bands to determine the power for the full-scale boat. Parsons intended originally to employ a single turbine and propeller, but ran into serious trouble with propeller cavitation. The final arrangement consisted of three turbines, each driving a separate shaft with three propellers on each shaft. The unit developed an incredible 2000 hp and in an attempt to publicise the use of the steam turbine in ships, Parsons decided to demonstrate the 'Turbinia' at the review of the British Fleet in the Solent on the occasion of Queen Victoria's Diamond Jubilee Celebrations in 1897. He created a sensation when he raced the first turbine vessel at the unheard of speed of 35 knots among the Fleet. The Admiralty was obviously impressed and Parsons was rewarded with an order for turbines for a new destroyer.

The main problem in applying turbines to ship propulsion was the high speed of rotation which created difficulties in propeller design, the problem eventually being solved by using reduction gearing in which two trains of gears were usually necessary, Parsons with his characteristic thoroughness even bought an old ship to use as a floating test-bed for geared turbines. By 1906, turbines were fitted in the Dreadnoughts and later in the liners Lusitania and Mauretania which, with her 70 000 hp turbines, remained in service until 1935.

As well as designing turbines, Parsons also turned his attention to the design of gears and developed an improved method of gear-cutting. He was involved in the design of servo-operated governing of turbines for land use. In 1925, the firm of C. A. Parsons took over the firm of Sir Howard Grubb and Company Ltd, and entered the telescope building business, an interest no doubt inspired by Parsons' father Lord Rosse.

In 1927, Charlie Parsons, as he is affectionately referred to on Tyneside, became the first engineer to receive the Order of Merit. Although Parsons possessed great mathematical ability he tried to use as little mathematics as possible in solving engineering problems and was not afraid to 'think with his hands'. He refused to accept conventional methods and was an original thinker of the first order.

On 11 February 1931, he died while cruising off Kingston, Jamaica.

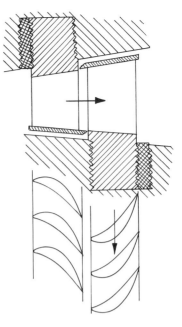

Parsons' turbine blades

Blaise Pascal 1623-1662

French physicist and philosopher

Blaise Pascal was born at Clermont Ferrand, France, on 19 June 1623 son of Étienne Pascal who was at one time President of the Court of Aids in Clermont. His mother died when he was only four years old and one of his sisters, Jaqueline, who later became a nun, helped to raise him and played a dominant part in his life. When Blaise was eight years old, the family moved to Paris but unfortunately they fell foul of the notorious Cardinal Richelieu when they went through a rather dangerous and difficult period. Young Pascal was tutored by various members of his highly talented family and, being a prodigy, worked with a feverish intensity which did nothing to improve his already failing health.

Although Pascal is regarded primarily as a philosopher, at least half of his writings were treatises on mathematics and physics. He astounded all with his knowledge of geometry which he had studied from the age of twelve. At sixteen he had proved one of the most important theorems of the projections of conics, known as 'Pascal's Mystic Hexagram' and he used this to derive over 400 propositions.

In 1641, Pascal's family moved to Rouen where he invented and patented a calculating machine using toothed wheels which could be used for addition and subtraction. A few years later, the whole family began to support the cause of Jansenism, a heresy violently opposed by the Jesuits. The heresy was based upon the teachings of St Augustine as interpreted by Cornelius Jansen, Bishop of Ypres, who maintained the perverseness and inability for good of the natural human will. Its supporters denounced the powerful Jesuits as hypocrites, a highly dangerous action in those days.

Shortly after the Pascals returned to Paris in 1650, Blaise's father Étienne died and Jaqueline entered the Convent of Port Royal, a centre of the Jansenist movement. Blaise joined the Jansenists and when their leader Antoine Arnaud was expelled from the Sorbonne for his heretical activities, his great friend Blaise Pascal wrote the first of his famous philosophical works, *Lettres Provinciales* in his defence. This great work greatly influenced the famous dramatist Voltaire. His best known literary work, *Pensées* was published unfinished, in 1670, eight years after his death.

Pascal is known generally as a great philosopher and theologian, but his work in science and mathematics was both original and outstanding particularly for someone who had received no conventional scientific education. In the field of mathematics, he was interested in the theory of probability and his work on the cycloid and solids of revolution acted as a stimulus to the development of the Calculus.

Following upon the researches of Evangelista Toricelli, he put the finishing touches to the theory of hydrostatics and showed that, in a fluid at rest, the pressure at a point acts equally in all directions. This is known to this day as 'Pascal's Principle' and is vital to the whole study of hydrostatics. Like Toricelli, Pascal experimented with the barometer and was the first to demonstrate the variation of atmospheric pressure with height. His experiments with barometers on the Puys de Dôme, a mountain near Clermont, are legendary. In a letter to his brother-in-law whom he asked for help in his experiment on the Puys de Dôme, he wrote, "–it is certain that there is more air at the foot of the mountain than at the summit and one cannot well say that nature abhors a vacuum more at the foot of the mountain than at the summit." Pascal's work on hydrostatics led to the invention of the modern barometer, the syringe and the hydraulic press of Joseph Bramah, which was not to come for another hundred and thirty years.

In 1658 Pascal's health began to decline rapidly and he died in 1662 at the early age of 39. The name of Pascal is honoured in the SI unit of fluid pressure, the 'pascal' which is equivalent to a force of one newton acting on a square metre.

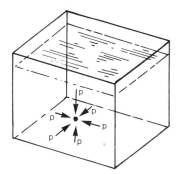

Pascal's Law — 'The pressure acting at a point in a fluid is the same in all directions'

Lester Allen Pelton 1829-1918

American inventor

In the fluid mechanics laboratories of engineering colleges, there is almost certain to be a small version of the impulse water-turbine known as a Pelton Wheel used for student experiments.

The invention of this mechanically simple yet highly effective turbine was one of those exciting moments in the history of technology when a keen observer instantly hits upon a brilliant idea which revolutionises a field of engineering. The story of the invention is an interesting one.

In 1848, gold was discovered in California and thousands of gold-mad prospectors dug and panned for the precious metal. One of them was a twenty-year-old carpenter from Vermillion, Ohio called Lester Allen Pelton. In later years, mining was carried out on a larger scale using hydraulic mining in which powerful jets of water washed the gold out of the alluvial deposits. As the workings were exhausted the water jets were used to drive crude water wheels to obtain power for mining machinery. The wheels, which were usually made of wood, had flat vanes working on the undershot principle, but these were later replaced by hemispherical cups with the jet striking at the centre.

One of the men tending the wheels was Pelton who had been unsuccessful at gold mining and had taken up mining engineering. One day he noticed that one of the wheels had moved along the shaft on a slack key and that the water jet was impinging on the inside edge of the bucket instead of at the centre. To his surprise the wheel was rotating at a much higher speed than normally with a consequent increase in power. Pelton built a wheel with the new configuration and later, when working at Camptonville in California on the construction of stamp mills, he introduced the split bucket which eliminated side thrust on the shaft. He spent the winter

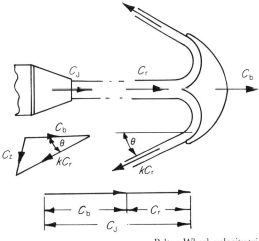

Pelton Wheel–velocity triangles

of 1878 endeavouring to improve the efficiency of his turbine using hand-beaten buckets made from empty oyster cans.

In the following year, Pelton made a model which he tested in the laboratories of the University of California with such great success that he received a prize for it. He took out a patent in 1880 and the rights were later sold to a company for which he worked as a consultant. In 1895, an extensive report was published on the Pelton Turbine by the Franklin Institute which awarded the inventor the Elliott Cresson Medal for his contribution to science in America.

By 1890 a 7 ft diameter Pelton Wheel was in use in an Alaskan mine to drive numerous stamp mills and ore crushers; it utilised a head of 400 ft and produced about 500 hp. Another machine at that time, running with a head of 2100 ft, realised an efficiency of over 80%. By the time of his death in 1918 Pelton's turbines were in use throughout the world for hydro-electric power schemes developing up to several thousand horsepower per unit at efficiencies of over 90 per cent with little change in the original design.

Henri Pitot 1695-1771

French hydraulics engineer and civil engineer

The Pitot tube, a simple but highly effective device for measuring the velocity of a stream of fluid, is the only achievement for which Henri Pitot the French engineer is generally known.

Henri Pitot was born on 3 May 1695 in the town of Aramon, between Avignon and Nîmes in the Languedoc district of France. His wealthy patrician parents, Antoine and Jeanne Julien Pitot could have afforded to give him an expensive education, but for some reason he stubbornly refused to be educated. He found the very idea of learning repugnant and after some years his despairing parents packed him off to the nearest military college so that he could pursue a career in the army. One day he saw a book on geometry in a shop window in Grenoble and decided to buy it. He found the contents so interesting that he returned home at once and spent the next three years studying geometry and other branches of mathematics.

Pitot then decided to go to Paris without any definite plans in mind but was fortunate enough to make the acquaintance of Réaumur, the great physicist famous for his temperature scale. Réaumur took an instant liking to the young man and gave him some good advice on improving his scientific education. He allowed him to use his private library and introduced him to the Académie des Sciences where he met other eminent scientists. The association between Pitot and Réaumur was to last over 20 years.

In 1723 Pitot became Réaumur's assistant in the chemistry laboratory in the Académie but he still kept up his mathematical studies. He received an invitation in 1740 from the Estates General of Languedoc to drain some swamps in the province. Pitot was apparently successful in doing this for he was next appointed Public Works Director for one of the three districts of the province and made Superintendent of the Canal Midi. He lived in Montpellier where he married Marie Léonine de Saballoua and they had one son who later became the Attorney-general of Montpellier.

Henri Pitot was an engineer who combined theory with practice, he wrote numerous papers on astronomy, geometry and mechanics, especially fluid mechanics and gave competent solutions to many problems in hydraulics, none of them requiring very advanced mathematics. His one book, *La Théorie de la Manoeuvres de Vaisseau* published in 1731 gained him admission to the Royal Society in London. His practical work included a road bridge attached to the famous Roman aqueduct, the Pont du Gard at Nîmes, and an aqueduct to carry drinking water to Montpellier.

The invention which immortalised the name of Henri Pitot is described in the *Mémoires de l'Académie Royale des Sciences*, entitled, 'Description d'une machine pour mesurer la vitesse des eaus courantes et le sillage des vaisseau.' It consists simply of a piece of glass tube bent at right angles which, when placed in a stream of water with one leg pointing upstream and the other vertical, indicates the velocity of flow

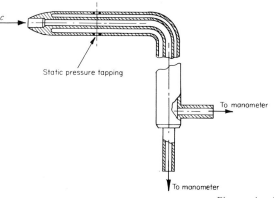

Pitot-static tube

by the rise of water in the vertical leg. Pitot realised that the height of water thus observed was the same as the height through which water would have to fall to achieve the same velocity V, that is $V = \sqrt{(2gH)}$, where H is the height in the vertical leg and g is gravitational acceleration.

This basically simple instrument has been elaborated and can be combined with a static pressure tube to give what is known as a Pitot-Static tube, used for the measurement of velocity in a closed pipe or duct. By connecting these tubes to a manometer or other pressure measuring device, the velocity of a fluid stream can be determined.

Pitot retired from engineering in 1756 and left Montpellier to return to his native town of Aramon where he died on 27 December 1771.

Max Carl Ernst Ludwig Planck 1858-1947

German physicist

The name of Max Planck, the great German physicist, who first proposed the quantum theory, appears in physics textbooks in connection with the well-known Planck's constant, h. It also crops up in Planck's Radiation Formula in the study of Heat Transfer.

Max Carl Ernst Ludwig Planck was born in the German city of Kiel on 23 April 1858, the fourth child of Johann Julius Wilhelm von Planck of Göttingen, a professor of Law who was descended from a long line of scholars, clergymen, lawyers and public servants. Max entered the Munich Gymnasium in 1867 and, as was the custom in those days, studied a wide range of subjects including music, literature and science. He chose science as a career although he was gifted in music and could play the organ and piano with great proficiency as a child. He was to retain a great interest in music for the rest of his life.

In 1874, Planck graduated and moved to Berlin where he worked under that brilliant pair of physicists Kirchhoff and Helmholtz and also came under the influence of the work of Rudolf Clausius on thermodynamics. In his Doctor's thesis, Planck criticised the definition of thermodynamic irreversibility put forward by Clausius.

His first university appointment in 1885 was as Extraordinary Professor of Theoretical Physics at Kiel, and on the death of Kirchhoff he was appointed to the same position at Berlin. He was promoted to Ordinary Professor in 1892 and held that post until his retirement at the age of 70. In 1918, Planck received the Nobel Prize for his brilliant work in physics, particularly in connection with the quantum theory. In 1926, he became a foreign member of the Royal Society.

Early in his career, Planck published a number of papers on a variety of scientific topics including several on thermodynamics, the dissociation of gases, osmotic pressure and chemical equilibrium. His great interest in the radiation of heat resulted from the work of Kirchhoff on 'black-body' radiation, where he had shown that the equilibrium temperature inside a cavity depended only upon the temperature and not upon the material of the cavity. Intrigued by this extraordinary result, Planck embarked upon a series of investigations which began in 1896 and lasted for five years. His efforts were rewarded; using W. Wien's radiation law and the formula of Rayleigh and Jeans, he obtained a new formula which agreed very closely with experimental results. Planck, however, was not satisfied until he had arrived at a suitable physical explanation of his findings and this he did by introducing his famous 'quantum theory' for radiation.

The theory asserts that electromagnetic radiation is emitted, not in a continuous wave motion, but in the form of 'packets' or 'quanta'. He showed that the energy was proportional to the frequency v, of the radiation, that is, equal to hv, h being Planck's Constant. In 1905, Albert Einstein proposed his theory in which he stated that monochromatic light consisted of particles called 'photons'. Neither Einstein nor Planck could accept the quantum theory as anything more than a step towards a more satisfactory explanation. Planck worked on a compromise between wave and corpuscular concepts until well into his eighties.

Max Planck's long life was clouded by a series of tragic events. His wife died at an early age in 1909, and in 1916 his son Karl died in action in World War I. In 1917 and 1919 two of his daughters died in childbirth. When the Nazis came to power, Planck pleaded unsuccessfully with Hitler for the release of several of his colleagues who were Jews and as a

Planck's formula: $E_v = \dfrac{hv^5}{c^3 [\exp(\frac{hv}{kT}) - 1]}$

E_v = Energy density
h = Planck's constant
k = Boltzmann's constant
T = Temperature
c = Velocity of light
v = Frequency

result he was forced to resign from the presidency of the Max Planck Society in 1937. Tragedy struck again in World War II, when first his eldest son was killed in action and then another son, Erwin was suspected of complicity in the plot to kill Hitler in July 1944, and was executed. Towards the end of the War Planck's house was destroyed by Allied bombing and most of his manuscripts were lost. In the confusion that followed he was rescued by the US Army.

The last two and a half years of Planck's life were spent peacefully in Göttingen where he witnessed the founding of the Max Planck Gesellschaft zur Forderung des Wissenschaften. He died aged 89 on 3 October 1947 at Göttingen.

Max Planck was not only a great scientist but also a philosopher and a man with a deep religious conviction which greatly helped him to bear his sorrows with fortitude. A churchwarden from 1920 until his death, Planck believed in an almighty, omniscient and benificent Creator who was, in his own words, "identical in character with the power of natural law."

Jean Louis Marie Poiseuille 1799-1869

French physician and physiologist

Jean Louis Marie Poiseuille was a rare example of a medical man who became famous in the world of engineering for his contribution to the field of fluid mechanics. As a result of his investigations into the flow of blood in veins and arteries, he developed an equation for the flow of a viscous fluid in a circular pipe which is also useful in many fields of engineering. The student of fluid mechanics will be familiar with Poiseuille's Experiment to find the viscosity of liquids using the simplest of apparatus, a large bottle of the liquid, a piece of capillary tubing through which the liquid flows into a measuring jar, and a stop-clock to time the flow. The equation governing the flow is due to Poiseuille but was discovered independently by the German engineer Gotthilf Hagen and therefore is known also as the Hagen-Poiseuille Equation.

Poiseuille was born on 22 April 1799 in Paris, and after studying medicine obtained his degree in 1828. He practised in Paris and became very interested in the mechanism governing the flow of blood in the veins and arteries and the pumping power of the heart. From experiments with various liquids in small-bore pipes, Poiseuille found that if the fluid were sufficiently viscous, the flow depended directly upon the pressure difference. He also found that the rate of flow was inversely proportional to the viscosity. Incidentally this was 50 years before Osborne Reynolds found experimentally that there existed two types of flow, laminar, or viscous flow used in Poiseuille's experiments, and turbulent flow where viscous effects were less predominant.

These experiments were all carried out in a most meticulous manner with excellent apparatus and sophisticated techniques and appeared in the Comptes' Rendus of the Académie des Sciences for 1841 under the title – 'Recherches expérimentales sur le mouvement des liquides dans les tubes de très petit diametre.' Poiseuille was the first scientist to use the mercury manometer for pressure measurement, a form is still used today for the measurement of blood pressure, the sphygmomanometer.

Poiseuille's name was given to the cgs unit of viscosity, the 'poise' with the subdivision 'centipoise'. The new SI unit, the 'Newton metre per second', which has not been named has replaced the 'poise', hence the famous name has unfortunately disappeared as an engineering term.

Poiseuille died in Paris on 26 December 1869.

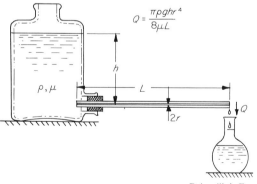

Poiseuille's Experiment

Simeon Denis Poisson 1781-1840

French mathematician and scientist

All engineers will be familiar with Poisson's Ratio, the ratio of the lateral contraction of a bar in tension to the longitudinal extension. This ratio, denoted usually by the Greek symbol v, is of vital importance in the theory of elasticity. They may also have come across Poisson Distribution used in Statistics and particularly applicable to most nuclear processes. Of interest to mathematicians is Poisson's Equation but Poisson is also renowned for his fine work in theoretical electricity and magnetism.

Simeon Denis Poisson was born on 21 June 1781 in the town of Pithiviers, Loiret, France. His family was very poor and by the age of fifteen he could do little more than read and write. Therefore he was sent to an uncle in Fontainebleau who arranged for him to take mathematics classes. The French Revolution was over and Simeon's father had been made head of the local commune and in that position was automatically sent a copy of the journal of the École Polytechnique in Paris which contained mathematical and scientific papers. Young Simeon began to read these and became intensely interested in mathematics. He took the entrance examination for the École Polytechnique and was successful. In 1798 he graduated and was top in mathematics, thus attracting the attention of Lagrange, renowned for his prowess in mechanics. Poisson also came into contact with other famous people, such as Fourier and Laplace, and began to carry out research into applied mathematics.

The theory of elasticity proved to have a great attraction for Poisson and he was interested in the molecular structure of elastic materials. He did useful work on the vibration of bars and plates and discovered the phenomenon of 'Poisson's Ratio'. This ratio lies between 0.2 and 0.4 for most materials and is about 0.3 for steel. It also relates Young's Modulus and the Modulus of Rigidity. In his book, *Traité de Mécanique*, Poisson introduced the trigonometric series for the first time and applied it to the investigation of the deflection curves of loaded bars.

Poisson also developed a theory of electricity in 1812, based upon gravitational theory and Lagrange's potential function, which agreed very closely with the experimental results of Coulomb. In 1824 he gave the world a wonderfully complete theory of magnetism and also investigated the phenomenon of induced magnetism.

Of course Poisson was fundamentally a mathematician rather than a scientist or engineer and he contributed significantly to topics such as calculus, differential geometry and probability; he also carried out important work on capillarity, heat, and astronomy.

This great mathematician and contributor to the theory of elasticity died in Sceaux, France on 25 April 1840.

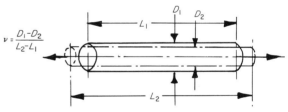

Poisson's Ratio

Ludwig Prandtl 1875-1953

German physicist and aerodynamicist

Reynolds, Froude, Mach, Schmidt, Graetz, Stanton, Grashof, Nusselt and Prandtl all have 'numbers' associated with their names. The student of fluid mechanics and heat transfer is faced with this seemingly endless succession of so-called 'Dimensionless Quantities'. Each number consists of several variables such as velocity, diameter, viscosity, conductivity, etc. arranged so that the group has no dimensions. The last one mentioned, is named after Ludwig Prandtl one of the most famous of them all, and his number is used extensively in the field of heat transfer for problems on heating and cooling by convection. It incorporates the fluid properties of viscosity, specific heat and conductivity.

The greatest contribution of Prandtl to fluid mechanics however is his hypothesis relating to the existence of a 'boundary layer' on surfaces over which a fluid flows. This hypothesis, which he introduced in 1904, bridged the immense gulf between empirical hydraulics and modern aerodynamics and hydrodynamics. It states that on the surface, or boundary over which a fluid flows, there exists a layer in which all viscous effects take place. Outside this

layer, which may be extremely thin for low-viscosity fluids, the flow is always non-rotational. For his work on boundary layer, the nature of turbulent flow and the separation of flow at the boundary layer, Prandtl has come to be regarded as one of the greatest contributors to the study of fluid mechanics. The application of his theories to the design of aerofoils and to flight in general has rightly earned him the title of 'father of aerodynamics'.

Ludwig Prandtl was born in the Bavarian city of Freising, now in West Germany, on 4 February 1875. He studied engineering and became Professor of Mechanics at the age of 26 in the University of Hanover where he set out to put the study of fluid mechanics on a sound mathematical basis. After four years at Hanover he moved to Göttingen, again as Professor of Applied Mechanics, and remained there until his death in 1953. At Göttingen, Prandtl established a world-famous school of aerodynamics and hydrodynamics and in 1925 he became Director of the Kaiser Wilhelm Institute (now the Max Planck Institute) for Fluid Mechanics.

Prandtl's discovery of the boundary layer in 1904 led to a better understanding of skin friction and drag on wings, struts, etc. and resulted in the streamlining of aircraft. His famous paper of 1918 the 'Theory of lifting surfaces', which followed similar work by the Englishman F. W. Lanchester in 1902-7, explained the process of airflow over the wings of aircraft and became known as the Lanchester-Prandtl Wing Theory.

This leading aerodynamicist did not confine himself solely to theoretical work but applied his theories to the airships and dirigibles of World War I and to the development of the monoplane as an advanced heavier-than-air machine. Prandtl showed how tests on model aircraft could be used to design full-scale machines; he also built a large wind tunnel at Göttingen, the very first to be used for the development of military aircraft during World War I. He did invaluable work on compressible and suspersonic flow and together with his development of wind tunnel testing this helped to pave the way for modern supersonic aircraft.

Like most great men of science and technology, Prandtl did not confine himself to one field of endeavour. He also contributed to the theory of elasticity and the strength of materials. In his study of the torsion of non-circular section bars, he introduced the 'soap film analogy' which has been so valuable to engineers. He investigated the large deflections of thin plates and the buckling of thin rectangular section struts. There is no doubt that his work greatly accelerated progress in the theory of elasticity. This great man was responsible for encouraging another giant in the field of aerodynamics, Théodore von Kármán, to take his Doctorate at Göttingen and work as his assistant on research into aircraft.

Ludwig Prandtl continued with his researches at Göttingen until his death on 15 August 1953.

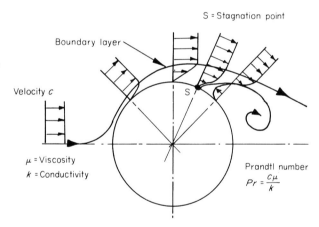

Flow separation from cylinder

Pythagoras of Samos c. 572 B.C. - c. 497 B.C.

Greek philosopher, astronomer, mathematician

Pythagoras is renowned mainly for his famous theorem in geometry concerning the right-angled triangle with which countless generations of children have struggled. He was however more than just a mathematician; he was a great thinker who founded a famous school of natural philosophy.

Pythagoras was born on the Greek Island of Samos about 572 B.C. and as a young man studied under the philosopher Anaximander and the great astronomer Thales, both of Miletus. He then spent years wandering throughout the Middle East gathering knowledge, until he finally settled down in Croton in Southern Italy where he founded the Pythagorean Brotherhood. This was a society which combined religious, scientific and political philosophies. Unfortunately none of Pythagoras' writings have survived so

that it is possible that he has been credited with some of the work of his disciples. It is generally agreed that he did invent the theory of numbers and put geometry on a sound footing with each proposition connected in a logical sequence. He derived proofs for a large number of Euclid's propositions, the most famous of course being that relating to the right-angled triangle which states that the square on the hypotenuse is equal to the sum of the squares on the other two sides. Pythagoras was familiar with the five regular solids known as the Platonic Solids and he and his followers realised that numbers such as $\sqrt{2}$ were irrational and could not be evaluated exactly. This knowledge was kept a close secret for fear of ridicule–it is said that one of their number was executed for disclosing the truth to an outsider.

Pythagoras was deeply interested in astronomy and he taught that the earth was a sphere and at the centre of a spherical universe. He analysed the motion of the sun, moon and the planets as the sums of two circular motions, a theory which persisted until the time of Kepler, twenty-two centuries later. His observations included noticing that the Morning Star called Phosphorus and the Evening Star known as Hesperus were both the planet Venus, and that the orbit of the Moon is inclined at an angle to the earth's equator.

The Pythagoreans asserted that all things were 'number' and that the truth of the physical world is contained in the mathematical relationships governing natural phenomena. Their philosophy included the doctrine of the immortality and the transmigration of the soul.

It is generally believed that Pythagoras discovered the simple ratios of the lengths of the strings of a lyre which give harmonious chords and there is of course a musical scale named after him. He and his disciples associated these lengths of lyre strings with the distances of the planets from the earth, in the belief that there was harmony in the spheres of the universe.

Pythagoras died *circa* 497 B.C. in the city of Metapontium in southern Italy.

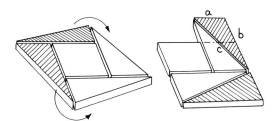

Pythagoras' Theorem $a^2 + b^2 = c^2$

William John Macquorn Rankine 1820-1872

Scottish engineer

A few months after the death of James Watt, another great Scottish engineer was born. He was William John Macquorn Rankine who, like Watt, was deeply interested in steam power. He was to give his name to the well-known Rankine Cycle upon which the cycles of all steam power plant, including nuclear plant, are based. Rankine is also equally well known in engineering for his work on Strengths of Materials particularly in connection with the buckling of struts where his name is linked with Gordon, to produce the Rankine-Gordon Formula.

William John Macquorn Rankine was born in Edinburgh on 5 July 1820 and, like many young men of promise in those days, was educated by his father, David Rankine, a retired army officer turned engineer. He tutored his son with characteristic Scottish thoroughness, and later William attended Ayr Academy and Glasgow High School and by the age of 14 had read and fully digested Newton's *Principia*, in the original Latin, and had studied physics to a high level. He graduated in Civil Engineering at Edinburgh University, winning a Gold Medal and in the process gaining valuable practical experience working for four years on the Dublin-Drogheda Railway in Ireland as a surveyor and construction engineer. He was also engaged on river and harbour works in Ireland. As Secretary of the Caledonian Railway, William's father was able to employ him as an engineer, and it was in 1842, at the age of 22 that Rankine published an extremely useful work on the form of railway wheels and metal fatigue in railway axles.

In 1855 Rankine was appointed Professor of Engineering at Glasgow University, a post he held until his death. At Glasgow he became involved in the development of the steam engine cycle and his brilliant book with the unpretentious title of *Manual of the Steam Engine*, introduced the world of theoretical thermodynamics to the practical engineer for the first time. The terminology and notation in the book is almost identical with that of modern textbooks on Applied Thermodynamics and Heat Engines. Rankine popularised the term 'energy' which had been introduced by Thomas Young half a century before, and he introduced a temperature scale which used an absolute zero and intervals equal to those of the Fahrenheit Scale. So that degrees Rankine or °R = °F + 460. This scale was in use in English-speaking countries until the adoption of SI units. It seems a great pity that Rankine's name will no longer be associated with a scale of temperature.

Students taking engineering thermodynamics will of course be familiar with the Rankine Cycle for a vapour and its variants based upon the theoretically optimum cycle of Sadi Carnot. The theoretical efficiency of such a cycle, the Rankine Efficiency, compares favourably with the Carnot Efficiency for the same temperature limits and may exceed 40 per cent for superheated steam.

In the field of Strengths of Materials, Rankine made a number of notable contributions, as shown in his *Manual of Applied Mechanics*, where he introduced the words 'stress' and 'strain' for the first time ever in English technical literature, in his discussion on 'stress at a point'. He also deals effectively with the theory of frames and struts and showed how the effect of shear should be included in calculating the deflections of beams and struts. In 1860 Rankine published an interesting paper on the forces acting on ships hulls due to waves.

It has been reported that Rankine was a very handsome and sociable man of gentle disposition with a great love of good music. His friends failed to understand why he never married, it may have been due to his passionate interest in engineering. He stands out as a theoretical man whose aim in life was to bring theoretical physics down from the dizzy heights to the level of the practical engineer.

Rankine gained considerable recognition for his wonderful contribution to science and engineering, which led to his being elected the first President of the Institution of Engineers in Scotland and the Fellowship of the Royal Society. He died in Glasgow on 24 December 1872, still in harness at Glasgow University.

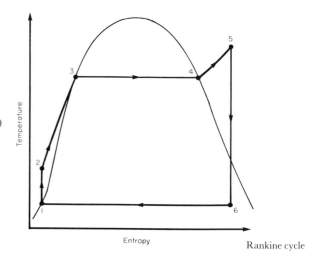

Rankine cycle

Lord Rayleigh (John William Strutt) 1842-1919

British mathematician and physicist

Although there are many examples of men who have been raised to the peerage for their outstanding work in science, one does not often find a peer by inheritance who devotes his entire life to scientific investigation. John William Strutt was such a man.

He was born at Langford Grove, Essex on 12 November 1842, the first son of the second Baron Rayleigh of Terling in Essex, England. John William attended both Eton and Harrow for short periods but had to leave because of ill health. He then spent a few years at a boarding school in Torquay where he showed little interest in the classics but was very good at mathematics. He eventually entered Trinity College, Cambridge at the age of twenty and studied mathematics under the famous Sir Gabriel Stokes, noted for his research on fluid mechanics. Stokes, who was fond of performing experiments in physics for his mathematics students, had a great influence upon Rayleigh, who graduated from Cambridge as Senior Wrangler in the Mathematics Tripos of 1865. He decided to follow a scientific career in spite of strong objections by members of his family, who considered him to be deserting his obligations to the title and estates.

Rayleigh's first paper on experimental work on galvanometers, was read to the British Association in 1868 and was the beginning of a flood of research papers. In 1871 he married Evelyn Balfour (the sister of James Balfour, destined to become the great statesman and Prime Minister of England) but a very serious bout of rheumatic fever shortly after his marriage nearly ended his career, however, a boat trip on the Nile seemed to cure him.

Strutt succeeded to the title in 1873 and moved into the ancestral home at Terling, setting up the laboratory which was to yield so much brilliant work. Visitors never ceased to be amazed at the simplicity of his apparatus which always gave such precise results. The death of James Clerk Maxwell in 1879 had left the Cavendish Professorship at Cambridge vacant and Rayleigh was asked by friends to stand for the post. The Agricultural Depression of the 1870's had considerably reduced the income from his estates and Rayleigh needed an extra source of income, so he applied for the post and was successful.

At that time the new Cavendish laboratories opened by Maxwell were short of apparatus and there was almost no practical instruction for science students so Lord Rayleigh made great efforts to introduce scientific apparatus and to produce an elaborate system of instruction. In 1884 after only five years at Cambridge, he resigned his Professorship and returned to Terling where he worked unceasingly until almost the day of his death, on a vast range of scientific topics. Rayleigh published 446 papers on researches into such diverse topics as sound, light, electricity, magnetism, vibrational mechanics, fluid mechanics, properties of gases. He received great public recognition, was awarded 13 honorary degrees, received the Order of Merit and was presented with a Nobel Prize in 1904. He also received awards from over 50 learned societies.

Lord Rayleigh had the good fortune to stay in excellent health until his death at Terling Place on 10 June 1919, where he left three recently completed but unpublished papers.

Out of the immense range of topics explored by Lord Rayleigh, engineers are concerned particularly with his work on vibrations and his theorem on Dimensional Analysis in fluid mechanics. The work on vibration is useful for shafts and contains his well known approximate method for many degrees of freedom. His Theorem on Dimensional Analysis paved the way for the use of similarity methods, so useful in fluid mechanics.

Rayleighs method

If x = function $(p, q, r, s, ...)$

where $x, p, q, r, s, ...$ are independent variables,

then x = constant, $p^\alpha . q^\beta . r^\gamma . s^\delta ....$

Osborne Reynolds 1842-1912

British scientist and mathematician

Osborne Reynolds was born in Belfast, Ireland on 23 August 1842, the son of the Reverend Osborne Reynolds, a Fellow of Queen's College Cambridge who had held several senior teaching posts. The obvious interest of his son in engineering was fostered by his father who himself had taken out a number of patents in connection with agricultural engineering. At nineteen Reynolds began work with a local mechanical engineer so that he could learn to be a mechanic in the two years before he was due to go to Cambridge. His plan was to obtain a comprehensive knowledge of practical engineering before taking Honours at Cambridge.

After graduating at the age of 26, Reynolds was elected to the newly instituted Chair of Engineering at Owen's College, Manchester, which was later to become the Victoria University of Manchester. In the same year he married, but tragically his wife died within a year. He was to marry again thirteen years later and have three sons and a daughter.

In his thirty-seven years at Manchester, Reynolds carried out investigations into numerous aspects of physics and engineering, including the properties of mixtures of gases and liquids, cavitation in hydraulic machines, dynamic stresses, thermodynamics and heat transfer. He was interested in the development of hydraulic pumps and turbines and in lubrication. It is important to note that Reynolds was both practical and theoretical in the field of hydraulics. In 1875 he took out a patent which suggested the use of fixed guide vanes in centrifugal pumps, the diverging passages formed by these vanes assisting in converting kinetic energy to pressure energy in the casing; this is a feature of many modern pumps. The first pump of this type was made and tested in the laboratories at Manchester. Reynolds also designed the first multi-stage centrifugal pump. William Froude's hydraulic brake was much improved upon by Reynolds and used by him to redetermine the mechanical equivalent of heat. In this connection he built an experimental steam engine and even anticipated the subsequent work of C. A. Parsons on steam turbines.

In 1886 Reynolds became the first to carry out a mathematical investigation of the hydrodynamics of oil-film bearings and showed that the load on the bearing is actually supported on a high-pressure wedge of oil. This work followed the famous experiments of Beauchamp Tower (1845-1904), who, in 1883 discovered the existence of a high oil pressure in a journal bearing. The work of Reynolds led directly to the development of the tilting pad bearing of the Australian engineer A. G. M. Michell in 1899 called the Michell Thrust Bearing, and to the whole subject of bearing theory and design. Reynolds also developed the theory of rolling friction upon which the design of ball and roller bearings depends.

During his long and distinguished career, Osborne Reynolds came into contact with many famous scientists and engineers. He first met the great J. P. Joule in Manchester in 1869 and they formed a lifelong friendship. Joule greatly admired the brilliant work of Rankine and corresponded regularly with that other eminent hydrodynamicist, Gabriel Stokes.

Osborne Reynolds died at Watchet in Somerset on 21 February 1912 at the age of 69. He was probably the most influential and successful theoretical mechanical engineer of his age, who combined great mechanical intuition with a brilliant flair for mathematics. He is justly honoured in having his name connected with one of the most important 'non-dimensional' quantities, the 'Reynold's Number'. This number and the associated notion of the two types of fluid flow, laminar and turbulent, have played a vital role in the study of viscous fluids.

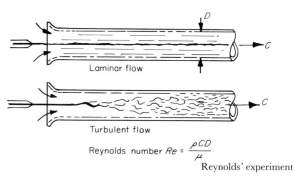

Reynolds' experiment

Count Rumford (Benjamin Thompson) 1753-1814

American soldier, statesman, scientist

In the year 1798, an American soldier of fortune, who had been knighted by King George the Third of England, was supervising the boring of cannon for the Elector of Bavaria when he noticed that a considerable amount of heat was generated. He bored a cannon in a tank of water over a long period of time and to the amazement of onlookers the water eventually boiled. The soldier, Benjamin Thompson, communicated the results of his observations to the Royal Society in London who, it is sad to say, took very little notice. It was not until 1842, 44 years later, that James Prescott Joule began his experiments to prove conclusively that heat was a form of energy and not a material substance.

Benjamin Thompson was born in the village of Woburn in Massachusetts near Boston on 26 March 1753. At the age of thirteen he was apprenticed to a merchant but he gave up his job when he was injured rather seriously in a firework display. After a brief period of study at Harvard, he worked initially in commerce, then as a teacher. When he was nineteen the impecunious young man married Sarah Rolfe who was fourteen years his senior and the widow of a wealthy landowner. Ben Thompson thus gained access to a social circle of which the Governor was a member. The Governor found him a commission as a Major in the Colonial Militia. Benjamin Thompson was of great help to the British in rounding up Army deserters, and at one time he did a little spying on the Rebel Army. These activities did not make him very popular with the American colonists and enraged by his treachery, a group of hotheads set out for his mansion to tar and feather him. He borrowed his brother-in-law's horse and galloped off into the night never to return.

When the military position of the British Army became untenable, Thompson decided it was time he left the country, so he deserted his wife and child and set off for London, ostensibly to report on the American situation. Such was his charm and personality that he made a great impression and rose rapidly in Government circles. Within four years he became an Under Secretary of State but still found time to indulge in scientific research. He constructed a ballistic pendulum to measure the velocity of musket bullets and hence calculate the power of gunpowder. It was then that he began to speculate on the nature of heat. He sailed with the British Fleet for three months studying ballistics and reporting on the gross inefficiency of the Fleet's tactics.

Thompson was eventually forced to leave England and to give up a promising career following strong rumours of his association with a notorious French spy LaMotte who, when caught had secret Naval documents on his person. He returned to America to take up a command in the King's Dragoons just as the War of Independence was ending. As a Colonel and a member of the British Military establishment, he was sent to the Continent as an aide-de-camp to the Elector of Bavaria having first been knighted by George III of England.

In Bavaria, Thompson carried out the greater part of his work on the nature of heat with unlimited financial assistance and no lack of trained artisans and mechanics, thanks to his position at Court. In his classic paper to *Philosophical Transactions*, he wrote, "What is heat? Is there any such thing as an 'igneous fluid'? Is there anything that with propriety could be called 'caloric'? . . . we must not forget to consider . . . that the source of heat generated by 'friction' . . . appeared evidently to be 'inexhaustible'! . . . it is difficult if not impossible to form an idea of anything . . . except it be motion responsible!"

In 1793, one year after the death of his estranged wife, Thompson was made a Count of the Holy Roman Empire for his services to the Bavarian Army and the country's social and educational system. Since his rise to fame began in the village of Rumford, the old name for Concord in Massachusetts, he took the title Count Rumford. His work included improvements in agriculture and the provision of work and sustenance for the poor and in Munich he organised 'Houses of Industry' for the destitute in which the kitchens were fitted with 'Rumford Boilers' and 'Rumford Roasters'.

As a result of jealousy in the Bavarian Court, Count Rumford fell from favour and was sent back to England as Minister Plenipotentiary to the Court of St James. Unfortunately the Elector had omitted to ask the permission of George III to make the appointment and the King refused to accept it. After an unsuccessful attempt to obtain a military position in America, Rumford applied himself to research once again and also to social work. He was a great philanthropist and was elected to the committee of 'The Society for Bettering the Conditions and Increasing the Comforts of the Poor.' The committee membership included William Wilberforce, the Bishop of Durham, and twenty-one MPs under the patronage of the King. Rumford was interested particularly in fuel economy and produced models of 'improved fireplaces, kitchens, flues and louvres for supplying rooms with tepid or fresh air and producing considerable savings in food and fuel consumed in cottages and public establishments.' In 1800 he founded the Royal Institution but after a few years his relationship with the directors deteriorated and he went off in a fit of pique to France. The country was then at war with England but that did not deter Rumford from settling in Paris in 1805 and marrying the wealthy widow of the great chemist Lavoisier who had died on the guillotine eleven years before. He established a laboratory in Paris and began to carry out his research once more but was immediately involved in bitter battles with the leading French scientists and with his wife to whom he referred in a letter as a 'female dragon'.

Not very long afterwards Rumford was separated from his second wife and he died suddenly a few years later on 21 August 1814 at Auteuil, Paris, leaving the residue of his estate to Harvard College. His grave is tended to this day by Harvard University.

Benjamin Thompson, Count Rumford was perhaps one of the most interesting and intriguing men in the history of science as well as being a prolific inventor and brilliant experimenter. The fact that he is relatively unknown is probably due to his inability to get on with people, although he was interested in solving social problems. This great inventor, engineer, sociologist and soldier of fortune contributed significantly to the founding of modern science.

Rumford's field coffee maker

Ernst Werner von Siemens 1816 1892

German electrical engineer and inventor

Werner Siemens was the eldest and most versatile of four brothers who were all gifted engineers and inventors. He was born at Lenthe, Hanover, Germany on 13 December 1816, the son of a farmer who himself was the youngest of fifteen children. The family originated in the old German town of Goslar and could trace it's origins back to 1538. Although they were chiefly agriculturalists, they ran oil and corn mills and mines and some had even taken up the learned professions.

Werner, the fourth of seven children, spent his first happy years at Lenthe. His father however, as a patriot who believed passionately in the unification of Germany, decided to move from the 'Royal British' province of Hanover to Meuzendorf in Mecklenberg-Strelitz. The young Siemens

went to school in Lübeck and then decided to study science and engineering. His father, who thought that the Berlin Academy would be too expensive, decided that the Army could provide just as good an education free. So one Easter the seventeen-year-old Werner set off on foot for Berlin to join the Prussian Army Engineer Corps. Unfortunately no vacancies existed so the disappointed boy went on to Magdeburg where against fierce competition he passed the entrance examination for the Artillery Corps. It required special permission from the King of Prussia for the 'foreigner' to be allowed to join the Army.

Siemens appreciated the strict Prussian discipline and studied mathematics, physics and chemistry with great enthusiasm. However, he still managed to get involved in a duel and, upon being imprisoned, set up apparatus in his cell. In 1835 he was transferred to the Engineers and Artillery School at Berlin where he acquired technical and scientific skill and in 1840 moved to Wittenburg, where he produced his first invention. This was the plating of articles with gold and silver. He wrote that it was one of the greatest moments in his life when he managed to coat a spoon made of German silver with shining gold. He sold the rights of the process for 40 louis d'or to finance further experiments including some on nickelplating. Siemens began to make money out of his inventions and, in 1843 sent one of his brothers, Karl Wilhelm, to England to develop his plating inventions also others related to electricity. His brother eventually returned to Germany but soon made his way back to England which became his second home. In England he ran the Siemens enterprises including the regeneration principle for waste heat recovery in furnaces.

In Berlin, Werner Siemens looked after the German activities and helped found the Physical Society. An important project was the development of Wheatstone's dial Telegraph in which he introduced automatic make-and-break circuits, and which was adopted by the German and Russian railways in 1846. In 1847 the world-wide firm of Siemens and Halske was founded in Berlin, J. G. Halske was Werner's brilliant mechanic. The factory made needle telegraphs, sounders for railways and gutta-percha insulated wire made in a machine invented by Siemens.

Siemens next turned his attentions to the generation of high power electricity. He developed his dynamoelectric generator which used electromagnets for the field instead of permanent magnets and invented the 'H' armature. He also investigated the possibility of electrically powered railways. Other projects included work on explosives and the rolling of steel.

It is impossible to discuss the life of Ernst Werner von Siemens without some reference to his brothers. Karl Wilhelm, seven years younger than Werner, started his career as an engineering apprentice and later went to England where his application of the regenerative process resulted in a reduction in the cost of producing pig iron by up to 20 per cent. He devised a scheme for using low-grade coal in the iron industry by employing a gas producer and developed the Siemens-Martin open-hearth steel process which was eventually to produce more steel than Henry Bessemer's converter. He designed the trans-Atlantic cable-laying ship 'Faraday' and built the electric railway at Portrush in Ireland. He was made an FRS in 1863, President of the British Association in 1882 and a year later was knighted, to become Sir Charles William Siemens.

Friedrich Siemens, who started his career as a seaman, helped Werner in his work and then went to England to sponsor Werner's telegraph system. His main interest however was in steel, helping Charles William. Karl Siemens looked after business interests in Russia where he introduced the telegraphy system. The news from the war in Crimea reached St Petersburg on the longest land-line of the day.

Ernst Werner von Siemens died at the age of 76 at Charlottenburg, Germany on 6 December 1892. He had experienced the satisfaction of seeing his great inventions come to fruition and attaining public acclaim worthy of his creativeness. It is interesting to realise that three of the magnificent team of brothers received titles, Werner in Germany, Charles William in England and Karl in Russia. Werner also has the honour to have an SI unit named after him. The 'siemen' denoted by 'S' is the new unit of conductance, the inverse of the ohm.

In his writings, Siemens made many statements of value to the prospective engineer. To students–"Don't specialise too early, one's views are narrowed. Professional work supplies the special knowledge!"

To the inventor–"Ideas have only limited value. The value of an invention lies in its practical realisation, the mental labour put into it and the money spent on it."
Also–"It's a long, hard road from a successful experiment to a successful mechanism, a road on which ninety-nine percent of inventions break their necks."

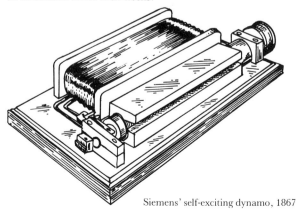

Siemens' self-exciting dynamo, 1867

George Stephenson 1781-1848

British engineer

George Stephenson, one of Britain's greatest engineers, was born on 9 June 1781 in a humble ivy-covered miner's cottage in the little village of Wylam about eight miles west of Newcastle-upon-Tyne. His father Robert was a coal-miner earning a weekly wage of 12 shillings, not very much on which to keep a family of six children.

George had practically no schooling and as a child earned a few pence tending cows until he was old enough to drive the horses working a hoist at the mine. He was immensely proud to be put in charge of the pumping engine at Water Row colliery where his father worked, a very responsible job for a lad of seventeen. His engineering education began when in his spare time he would clean and service the engine. He worked in several collieries and finally in Killingworth West Moor Pit in 1804. He began to learn to read and write at eighteen and at the age of twenty-one married Fanny Henderson the servant of the mineowner. His son Robert, who was to be a great engineer in his own right, was born in 1803.

In 1805 tragedy struck when George lost both his wife and daughter and his father was seriously injured in the pit. He was on the verge of emigrating to America as had several of his relations and friends, when he acquired a new interest in life, that of providing a good education for his son Robert. He scrimped and saved to send him to the High School in Newcastle and sat in the evenings helping the lad with his work.

Stephenson was eventually put in charge of all the machinery at Killingworth and made responsible for the engines of the whole group of collieries, acting as consultant for the owners. He became extremely competent at designing and making engines and made thirty-nine with powers up to 200 hp. He introduced iron rails for trucks with the important innovation of flanged wheels; the trucks at this stage were hauled by stationary engines. Among the many inventions connected with mining was Stephenson's famous 'Geordie Lamp' for which he was awarded £1000 by public subscription, but not without his being brought into some controversy with Sir Humphrey Davy who had invented the equally famous 'Davy Lamp'.

Because of the rising cost of fodder for horses, mine-owners were being forced to consider the use of steam engines which ran on rails. Many unsuccessful attempts were made including a high-pressure Trevethick-type locomotive built at Gateshead-on-Tyne which was to be used at Stephenson's birthplace Wylam. Two Trevethick loco's with rack and pinion drive were used near Leeds and two more on Tyneside. Stephenson's employers commissioned him to build a locomotive for Killingworth which he called the *'Blücher'*. It made its highly successful debut on 25 July 1814 and was followed by *'Wellington'* and *'My Lord'*. George helped to found a locomotive works near Newcastle which his son Robert, recently returned from the supervision of mines in Colombia, joined as manager in 1827. Stephenson's first railway of any size was opened at the Hetton mine near Sunderland in 1822 for which he made five locomotives popularly known as 'iron horses'. At the Newcastle works locomotives were made for the newly opened Stockton-Darlington railway, of great importance to the development of railways since it would result in the opening up of new markets for Durham coal; the first of these locomotives was named *'Locomotion'*.

In 1830 the first mainline passenger railway in the world was opened between Manchester and Liverpool with the world famous *'Rocket'* being adopted for the first run. The construction of this railway was undertaken after an Act of Parliament was passed following a tremendous struggle by the rough-spoken, uneducated miner's son against the smart London lawyers representing the landowners who tried in every way to thwart his plans.

In the previous year Stephenson's reputation was greatly enhanced by his success in the historic Rainhill Trials in which the prize was £500. The locomotives were to weigh less than six tons and haul a 20-ton load at a speed of at least ten miles per hour. Only four locomotives took part, including John Ericsson's *'Novelty'* which many thought would be the victor. The *'Rocket'* was modified to produce 20 hp with an unbelievably small fuel consumption of about 18 lb of coke per hp hour.

George Stephenson was to live to see railways spread throughout the whole of the Continent of Europe and to many other parts of the world. A typical story of the man is concerned with the multiplicity of railway gauges then in use.

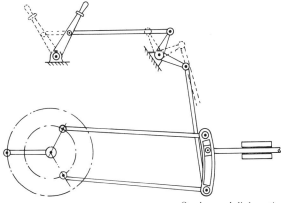

Stephenson's link motion

When locomotives were ordered from Stephenson's works for the first German railway between Nürnberg and Fürth in 1835, a special gauge was requested which was different from that used by George. He completely ignored the request and to this day all German railways use the British gauge of 4 ft 8½ in. Incidentally this gauge used initially in the mines in North East England determined the railway gauge for the whole world.

George continued to play a leading part in the development of the railways and eventually moved to an estate at Tapton near Chesterfield so that he could supervise what was to become the Midland Railway. He was saddened by the ensuing 'railway mania' in which large amounts of money were lost and he did his best to prevent it. After a short illness George Stephenson died at the age of 67 and was buried near his home at Tapton on 12 August 1848.

Although George Stephenson was by no means the originator of the steam locomotive, there is no doubt that his flair for engineering design coupled with a dogged determination to succeed was responsible for the rapid development of railways in Britain and eventually throughout the world. With the possible exception of his Link Motion there is little to connect Stephenson's name with a course in engineering; he is not responsible for any formula or principle. His achievements do however remain as a tremendous source of inspiration to the young engineer.

Robert Stirling 1790-1878

Scottish minister of religion and inventor

In 1816, a young, newly-ordained minister of the Church of Scotland filed a patent with the rather lengthy title, 'Improvements for Diminishing the Consumption of Fuel and in Particular an Engine Capable of being Applied to the Moving of Machinery on a Principle entirely New!'

The patent, taken out on 16 November (No. 4081), related to the Stirling Engine, a hot-air engine named after Robert Stirling who was born in Perthshire, Scotland in 1790. The engine operated on the well-known Stirling Cycle in which the working fluid was heated externally. It is interesting to note that the patent was also signed by James Stirling, Robert's younger brother. The patent drawings are obviously the work of a trained engineer and, although the notion of the inventor being a parson is more romantic, there is a strong possibility that James was the innovator. Two further patents on the engine were taken out in 1827 and 1840. Much has been written about the Stirling Engine recently as a possible solution to the pollution problem.

Although little seems to be known of the Stirling brothers, Robert is credited with the invention, but it was James who explained the working of an engine on test in 1845 to Sir George Cayley, Sir John Rennie, and Robert Stephenson who were present at the trials. It is possible that Robert conceived the idea and James carried out its development. Robert also designed and made scientific instruments.

The first Stirling engine to be put to work was a modest affair of 2 hp, used to pump water from a stone quarry in Scotland. The brothers introduced a number of improvements in the many engines they designed and built between 1820 and 1850. One of the largest of these produced 45 hp and was used in a Dundee foundry. All these engines operated on hot air produced by heating one end of the working cylinder by means of an external furnace. The Dundee engine required a temperature of over 600°F and not surprisingly the cylinder cover had to be replaced regularly.

The early engines produced by the Stirling brothers were subject to serious metallurgical problems because of the extremely high temperatures necessary to give adequate power. The famous Swedish engineer John Ericsson was fascinated by the principle of the Stirling engine and saw in it the successor to the steam engine. He built a number of Stirling hot air engines culminating in the huge engines installed in the ill-fated ship the *John Ericsson*. These unfortunately burned out and had to be replaced by steam engines.

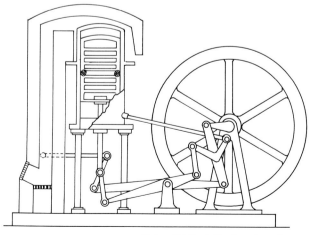

Robert Stirling's first hot-air engine

But the Stirling Cycle has always created great enthusiasm. One interesting application was that developed by Malone in 1931 who built an amazingly lightweight and efficient engine to replace the inefficient steam locomotive engine. It used for its working fluid extremely high pressure, high temperature water and although the prototype ran successfully it was never developed.

At the present day, over one and a half centuries later, legislation and economic pressure have resulted in a renewed interest in this quiet, low-pollution, high thermal efficiency engine. For at least thirty years the well-known firm of Philips has developed the Stirling engine and in 1958-60 produced an advanced design using helium as a working fluid, which realised an efficiency of 40 per cent. General Motors has spent a vast amount of money trying to develop the principle with designs for engines of up to 5000 hp and recently the United States Atomic Energy Commission and the National Heart Institute of America have collaborated in the production of a nuclear-powered aritificial heart operating on the same principle as the original Stirling Engine. It uses blood instead of water to cool the working fluid and the furnace is replaced by a small piece of radioactive isotope.

Robert Stirling must have found sufficient time to devote to his flock in spite of his engineering interests, for he was given the title 'Father of the Church of Scotland'.

In 1819 he married Jane Rankine, the daughter of a Galston wine merchant, and by her he had three sons, Patrick and William, who became engineers, and David who became minister of religion. After two years of failing health, Robert Stirling died on 6 June 1878 at Galston, Ayrshire, where he had lived for 53 years.

George Gabriel Stokes 1819-1903

British mathematician and physicist

The great era of physics at Cambridge began with the brilliant work of Gabriel Stokes in 1837. His first research was in the field of hydrodynamics, particularly in connection with fluid friction, and fluid mechanics students will be familiar with the experiment to establish Stoke's Law for a body falling in a viscous fluid. His other interests included the study of elasticity and he carried out experiments to determine the ratio of the elastic constants E and G.

George Gabriel Stokes was one of the large family of the Rector of Skreen, a village south of Sligo in Ireland. He was born on 13 August 1819 and received his early education from the parish clerk who taught him arithmetic. At the age of thirteen, he was sent to a church school in Dublin where his ingenious solutions to some geometrical problems attracted the attention of the mathematics master who was instrumental in getting him into Bristol College. At Bristol, Stokes won a prize for 'eminent proficiency in mathematics' and he was accepted for Pembroke College, Cambridge in 1837. He became Senior Wrangler and Smith's Prizeman, then in 1841 was elected Fellow of his College and later became Master. In 1849, Stokes was elected Lucasian Professor in Mathematics, a post he was to hold until his death.

Although Stokes was an eminent mathematician, he always emphasised the great importance of applying mathematics to physical problems. In addition to his work on hydrodynamics and elasticity, Stokes did valuable work on optics and introduced the word 'fluorescence' to science, a phenomenon which he discovered was produced by ultraviolet light.

Lord Rayleigh wrote that Stokes was a wonderful experimenter and that he achieved the most wonderful results with the most modest equipment, which he set up in a narrow passageway in his house. His annual lectures on physical optics were said to be a delight to attend because of the wonderful demonstrations and the fact that the work was fresh from his investigations. In 1845, Stokes became

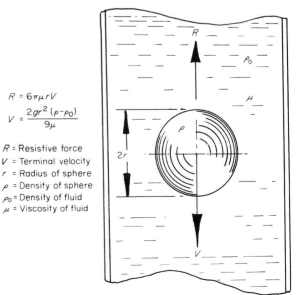

$R = 6\pi\mu r V$

$V = \dfrac{2gr^2(\rho-\rho_0)}{9\mu}$

R = Resistive force
V = Terminal velocity
r = Radius of sphere
ρ = Density of sphere
ρ_0 = Density of fluid
μ = Viscosity of fluid

Secretary of the Royal Society and his output of scientific work fell off considerably. He held the post for over thirty years and then served as President for five years.

George Gabriel Stokes died on 1 February 1903 at the age of 84, still a Professor of Mathematics. It is some measure of the respect in which he was held in the world of science that, on the occasion of the Golden Jubilee of his Professorship, many outstanding scientists from all over the world should visit Cambridge. It was said that his only fault was a tendency to be overcautious, but he was a good administrator and was Member of Parliament for Cambridge University for several years.

The name of Stokes will be remembered for his Theorem and for his famous Law. The name used formerly for the unit of kinematic viscosity was the 'stoke' with its subdivision the 'centistoke' but this has now been superseded by the SI unit.

Nikola Tesla 1856-1943

Yugoslavian/American electrical engineer

Nikola Tesla was one of the great inventive geniuses of the late 19th century in the field of electrical engineering. It is rather sad therefore that to many electrical engineers his name is associated solely with the SI unit of magnetic flux density, and not also with the major role that he played in the development of alternating current generation, distribution and utilisation.

Nikola was born in 1856, the youngest of five children whose father was the Rev Milutin Tesla, parson in the little town of Smiljan in the province of Lika in Serbia, then known as Croatia. Milutin was a tall handsome man, a fine speaker who knew the Bible by heart, his wife Djouka was an attractive woman who, although she could not read or write, spoke French, German and Italian as well as her native Serbian.

From an early age, 'Nikki' was inventing; he produced a water wheel to beat eggs for his mother, blow-guns and pop-guns; he tried to fly but broke three ribs and spent six weeks in bed. When the family moved to the larger town of Gospic he made a great impression by diving into the river to get the town's new fire pump to work, incidentally drenching the assembled dignitaries. To earn money to help him attend the Higher Real Gymnasium at Carlstadt he worked at the local library, having first taught himself French, German and Italian. At the gymnasium he studied mathematics and physics and became fanatically interested in the work of the great American electrical engineer, T. A. Edison. He even learned English so that he could study Edison's papers.

At the age of seventeen, Tesla was faced with the alternative of three years in the army or joining the church. Both prospects horrified him since he was eager to further his engineering education. He deliberately exposed himself to cholera—an epidemic was raging in Gospic at that time—caught the disease and was declared unsuitable for national service by his doctor. After a frustrating year's convalescence, he entered the Polytechnic at Graz, Austria where he showed outstanding ability in engineering. The Dean of the Technical Faculty, in a letter to Tesla's father, described him as a 'star of the first magnitude'. At the Polytechnic, during a demonstration of a Gramme motor, he amused the Professor and the class by suggesting the possibility of dispensing with the commutator with its inherent power loss. After leaving Graz, Tesla went to Prague University where he was able to finance himself from the money which he earned working in a factory. He studied English Literature, Advanced Mathematics and Advanced Electrical Engineering and obtained his diploma on the day his father died.

Tesla's first post was with the Central Telegraph Office in Budapest as a junior draughtsman on starvation wages, but by the time he was 25 he had been promoted to manager as a result of improvements he had made in telegraphy. It was at this time, while walking in the countryside, that he hit on the brilliant idea of polyphase alternating current and the use of a rotating magnetic field for ac machines which would dispense with the commutator. After short spells in France and Germany, he decided to emigrate to the United States and arrived in New York in 1884 with 5 cents and carrying a brown-paper parcel and a letter of introduction to his boyhood hero Thomas Alva Edison. He was somewhat dishevelled and bruised after being involved in a fight with mutineers on the ship.

Edison decided to employ Tesla who threw himself into his work with tremendous enthusiasm working incredibly long hours on new types of power transmission and generation. He developed a highly successful arc light for street lighting but, being inexperienced in money matters, was exploited and soon became penniless. He left Edison after a fight and even worked for a time as a labourer laying cables until he could raise money to start his own firm, the Tesla Electric Company, in which he developed ac power systems, generators and motors.

He also produced transformers to produce high-voltage and consequently low-loss transmission and in 1889 accepted an offer to merge with the Westinghouse Electric Company. Incidentally the founder of this firm, George Westinghouse, was from the aristocratic von Wistinghausen family from Westphalia. Tesla was offered a million dollars and royalties of a dollar per horsepower developed by his machines. Working in his own laboratories, he continued to produce many inventions including the famous 'Tesla Coil', a high-frequency, air-cored transformer capable of producing sparks 5 m long. It has since been used for therapeutic purposes but has little application elsewhere. He played a major role in harnessing the immense power of the Niagara Falls, thus fulfilling a boyhood dream.

Nikola Tesla never married and became something of a recluse in later life. He was obsessed with the idea of radio transmission and predicted many modern developments but his many experiments in the attempt to transmit power without wires never succeeded.

Money which he spent as fast as he made it, meant little to Tesla, and he was robbed of millions of dollars over patent infringements. This man, whom many undervalue, was offered many honours all of which he refused; he would not accept the Nobel Prize offered in 1917 because it was awarded jointly with Edison, and he adamantly refused to tour the USA and Europe giving talks and lectures.

In the early days of World War II, the ageing Tesla was extremely depressed by the evil uses to which his beloved science was being put and he was greatly saddened by the Nazi invasion of Yugoslavia. He lived in seclusion in hotels in New York, first in the luxury of the Waldorf Astoria, which he was forced to leave for not paying his bill, and then in less salubrious lodgings.

On 6 January 1943, a severe electrical storm was lashing New York. Tesla was standing at a window shaking his fists at the storm and shouting, "I have made better lightning than that!". He clutched at his heart and collapsed. A day later he died aged 87.

Nikola Tesla was one of that rare breed known as 'inventors', a word unfortunately not encouraged in these days of specialisation. Out of the vast flood of ideas to come from this somewhat eccentric genius many have made a tremendous impact in the field of electrical engineering even though others may appear to be ill-conceived.

In recent years Tesla has been honoured by having his name given to the SI unit of magnetic flux density, the 'tesla' being equal to one 'weber' per square metre and having the symbol 'T'.

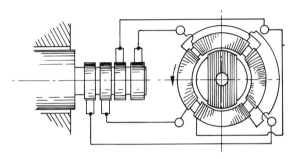

Tesla's first induction motor, 1887

Stepan Prokof'yevich Timoshenko 1878-1972

Russian/American scientist, engineer and teacher

Stephen P. Timoshenko is one of the most widely known authors of books on engineering, particularly in the field of Elasticity, Strength of Materials, and Structures. His books *Strength of Materials, Theory of Elasticity, Theory of Elastic Stability*, and *Theory of Plates and Shells*, are well known to engineering students and lecturers throughout the English-speaking world and have been translated into many languages. His *Advanced Dynamics* has been translated into Amharic, Arabic, Armenian, Bengali, Ewe, Fanti, French, Fula, Ga, Hausa, Igbo, Indonesian, Luganda, Malay, Persian, Peshtu, Serbo-Croat, Spanish, Swahili, Twi, Urdu, Wolof, and Yoruba. He has written five books in Russian and thirteen in English, thirty scientific papers in Russian and sixty-six in other languages.

Stepan (Stephen) Timoshenko's father, Prokop Timofeyevich worked as a surveyor, later assisting in the parcelling out of land to the peasants under the reforms of Alexander II. He married a Polish girl and on 23 December 1878 Stepan Prokof'yevich was born.

Young Timoshenko spent a happy childhood on the farm which his father worked and very early in his life decided to be an engineer, particularly on the railways. He had a poor early education but did well at the 'Realschule' at Romny particularly in mathematics, although he found algebra rather difficult. He actually found it necessary to learn Russian since he spoke Ukrainian. It was at Romny that Timoshenko discovered that he loved teaching, finding it a joy to explain the theories to other students.

After passing a competitive examination, Timoshenko was accepted for the course at The Institute of Engineers of Ways and Communications at St Petersburg. He considered the training in mathematics and mechanics very unsatisfactory, added to which there was considerable disorder due to student strikes and political unrest. He found himself on the side of political and social reform but against the suggestion of retaliation against property owners.

In 1901, Timoshenko graduated and took a job as an engineer on the railways. He served for a year in the Army and in August 1902 married Alexandra Archangelskaya whom he had met when she was a medical student in St. Petersburg. He then worked in the Mechanics Laboratory at the Institute of Ways and Communications carrying out tests on concrete and railway lines but found that there was practically no research being undertaken. He attended an extra mathematics course and then went to the St Petersburg Polytechnic Institute where he found conditions infinitely superior and the laboratory staff excellent. He began to learn English and was greatly impressed by Rayleigh's 'The Theory of Sound' finding the method for torsional vibrations on shafts particularly useful. At that time he published his first work, 'O Yavleniyakh Rezonansa v Valakh' (Resonance Phenomena in Shafts), using Rayleigh's Method.

Timoshenko then spent an interesting period studying under the great Ludwig Prandtl at Göttingen University in Germany, after which he began lecturing at Kiev University in 1906. His first ever lecture on Strengths of Materials, was delivered to over 400 students, and he admitted that he never got over his nervousness at the start of a lecture even though he always memorised his notes. In 1911, Timoshenko was one of three deans to be sacked from Kiev because of their objection to restrictions on the intake of Jewish students. Over 40 per cent of the staff resigned.

Out of work, Timoshenko made some money out of the publication of a book and received a prize of 2500 roubles associated with the Jourawski Medal. In 1912, he acted as consultant for the Russian shipyards, worked on his book on *Strengths of Materials* and continued with a little school teaching. He travelled to England where he met Rayleigh and saw the laboratories at Cambridge. He was amazed at the shortage of technicians and the untidy laboratories with their primitive home-made devices which compared badly with the highly organised German laboratories lavishly supplied with the best of equipment.

In 1913, Timoshenko went to St Petersburg as Professor at The Institute of Ways and Communications and later at the Electrical Engineering Institute, and in 1914 he finished his book *Theory of Elasticity*. War had broken out and by 1915 life in St Petersburg had become intolerable with riots and food shortages, and classes were too small to continue. He found himself once more in Kiev with fighting taking place between the Bolsheviks and the Ukrainians; the city was badly damaged by shelling and food was scarce. Under the Bolshevik occupation salaries were not paid.

In 1920 Timoshenko eventually managed to escape to Yugoslavia via Constantinople and settled in Zagreb with his family, but prospects were poor and in 1922 he accepted the offer of a job in USA with a firm that specialised in the balancing of machine rotating parts at a salary of 75 dollars per week. He decided to emigrate and settle in Philadelphia, leaving his eldest daughter Anna and his son Gregory in Berlin to receive a good engineering education in his opinion

not then possible in America. His employers went bankrupt and Timoshenko joined the famous firm of Westinghouse where he worked on stress analysis and photoelasticity. He met eminent men like J. P. Den Hartog and the Swedish engineer C. R. Soderberg. At Westinghouse he initiated evening postgraduate classes in applied elasticity and mechanics both of which the engineering graduates knew next to nothing.

In 1927, Timoshenko obtained US citizenship and was once again involved in teaching. He was offered and accepted the Chair of Research in Mechanics at the University of Michigan but was disenchanted with the American method of teaching engineering; he found that, instead of giving lectures, the staff gave work to the students to be learnt and that the system was one of 'interrogation rather than explanation'. He soon realised that the lecturers regarded him as a 'competitor' and not as a colleague, unlike the friendly attitude of the engineers at Westinghouse who would readily ask for advice. The quality of the professors, he thought, was inferior to that of his engineering friends, and he was amazed at how little influence he had over students and the shaping of courses.

To overcome the deficences in teaching of Strengths of Materials he organised a Ph.D. course on 'Thin Bars and Plates' and a summer school on Mechanics, on Vibration Theory and a Selection of Problems in Strength of Materials which was attended by over 50 teachers. He encouraged the use of outside lecturers which included R. V. Southwell, and his students included famous names like M. M. Frocht and J. N. Goodier. In 1935, Timoshenko was invited to run a course at the University of California at Berkley and found time to visit Von Kármán, the great aerodynamicist, at Pasadena.

Timoshenko liked California and in 1936 he joined the staff of Stanford University where he remained teaching until 1954 when he was 76 years old. In 1946, he received the James Watt Gold Metal in the same year in which his wife died. In 1952 he finished his *History of Strengths of Materials*. He continued to work on his books after retirement and also found time to return to his Russian homeland in 1958, where he visited many engineering teaching establishments. In 1959 he was elected to membership of the Soviet Academy of Sciences.

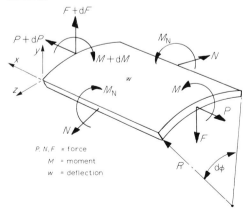

Element of stressed shell

Evangelista Toricelli 1608-1647

Italian physicist

Most barometers today are of the dial type but there are still many mercury column barometers to be seen, especially in laboratories where an accurate reading of atmospheric pressure is often required as a reference pressure in experiments. This type of barometer was invented three and a half centuries ago by an Italian mathematician and physicist called Evangelista Toricelli.

Toricelli, born in Faenza near Ravenna, in Northern Italy, on 15 October 1608 when the great Galileo was 44 years old, was educated at Rome University where he studied mathematics. After leaving university he experimented in science and published a book on his work which so impressed Galileo that he invited him to Florence to discuss his achievements. Toricelli accepted the invitiation and met the

Galileo's post of Court Mathematician under the patronage of the Duke of Tuscany was offered to Toricelli who gratefully accepted. He began to work on a number of scientific projects but was particularly interested in the then unexplained nature of a vacuum. It had been Galileo who had first suggested the problem but the solution to it was to make Toricelli famous.

It had always puzzled scientists as to how it was possible to lift water by means of a piston or by a siphon; the force acting to produce the upward motion was unexplained. Toricelli solved the problem when he suggested that air had weight and that like a liquid could exert a pressure in all directions, a fact that was later to be substantiated by Blaise Pascal. He showed that water could be lifted to a height of only about 10 metres by the pressure of the air at sea level. In 1643, he took a glass tube, 120 centimetres long closed at one end, and filled it with mercury; he then upended it in a mercury bath and thus produced the first barometer. The column of mercury was about 76 centimetres high with a space at the top which Toricelli realised must be free of air. He had created for the first time, an almost perfect vacuum, which is now referred to as a Toricellian Vacuum, and it was produced seven years before Otto von Guericke carried out his historic experiment in Magdeburg with two teams of horses trying to separate evacuated hemispheres.

Toricelli had observed that the height of the mercury in his barometer varied slightly from day to day showing that the pressure of the atmosphere must vary. A few years later Blaise Pascal followed up Toricelli's work by taking some barometers to the summit of the mountain Puys de Dôme near Clermont in France and found that there was a considerable fall in height of the mercury column. He had discovered that the pressure of the atmosphere fell with height.

Toricelli's findings came at a very crucial time in history, for they helped to pave the way for the development of the steam engine and the hydraulic press of the Industrial Revolution. Toricelli died on 25 October 1647 in Florence at only 39 years of age. He gave his name to a rather unusual unit of pressure, the Torr, equal to one millimetre of mercury.

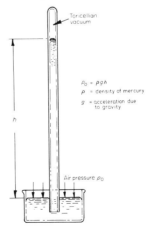

Barometer

Richard Trevethick 1771-1833

British engineer

Richard Trevethick was born on 13 April 1771 in the parish of Illogan, Cornwall, son of the mining engineer Richard Trevethick who already had four daughters. Shortly afterwards the family moved to Penponds where the children went to school until they were about eight years old.

When Richard was about four years old, his father, as manager of Dolcoath Mine, had installed a Newcomen steam engine. In 1777, the first low-pressure Watt engine was installed in Cornwall with a fuel consumption only a quarter to a third of that of Newcomen's engine. Mineowners were soon finding the Watt engines expensive to buy as the royalties were excessive and the patent so tightly drawn up by Boulton and Watt that engineers who could easily make engines much more cheaply were unable to do so without fear of litigation. Many Cornish engineers defied the patent and young Trevethick tried to negotiate with the men at Soho but without success.

In 1797, Richard Trevethick senior died and later that year his son married Jane Harvey who was to prove a wonderful companion, especially during times of financial difficulties to this tempestuous man with a tremendous capacity for work. Trevethick now began to work on a high-pressure steam engine. He argued that with a pressure of several atmospheres, instead of Watt's pressure of slightly over an atmosphere, the power would be increased greatly, and the condenser and air pump dispensed with. James Watt in characteristic frankness said that Trevethick ought to be hanged for using pressures which were murder for those attending the engine. Trevethick was bitter about Watt's comments but regarded them as somewhat of a challenge. He built an engine using a cylindrical boiler of cast iron a rather dangerous material to use, with the engine cylinder inside the boiler to keep it hot. The firebox was cylindrical with a return flue.

In the year 1800, Watt's patent expired and engineers everywhere tried to improve upon his design. Trevethick built a highly successful double-acting high pressure engine for Cook's Kitchen Mine and as a result his fame spread not only in Cornwall but also in South Wales, so that by 1804 he had sold about fifty engines of that type. He had already made a model locomotive at this stage and he proceeded to construct a full-scale machine which was tested on Christmas Eve 1801. The machine, known locally as 'Dick's Firedragon', negotiated a steep hill and then ran out of steam. The crew and helpers then pushed the machine into a shed and adjourned to a nearby inn to refresh themselves on roast goose and ale. Unfortunately, they forgot to draw the fire and all the water in the boiler evaporated so that the whole machine and the world's first garage went up in flames.

Other attempts to make locomotives to run on rails were unsuccessful due mainly to deficiencies in the rails. One locomotive called the 'Catch-me-who-can' was used as a demonstration model and ran on a small circular track at Euston near Russell Square. The public payed one shilling a time for a ride but the venture didn't pay and soon afterwards the discouraged Trevethick abandoned railways altogether. It was another eight years before Stephenson's first locomotive was completed. As a firm believer in the use of steam power, Trevethick built a steam dredger, a paddle steamer and a steam threshing machine; he also attempted to drive a tunnel under the Thames but this had to be abandoned. In 1810, he suffered a serious illness and in 1811 went bankrupt.

Back home in Cornwall, Trevethick turned once more to the building of mine pumping engines using high pressures and greater expansion of the steam. He improved his fire-tube boiler which came to be known as the Cornish Boiler and at the age of forty-two he embarked upon a remarkable adventure in which he was to spend ten years in Peru.

In Peru, there were rich silver mines high up in the Andes about 160 miles from the capital Lima, but they were a failure because of constant flooding. A representative sent to England realised that Watt's atmospheric engines would not be very efficient at 14 000 ft and would be too heavy to transport over mule tracks, so he bought a small Trevethick engine from a shop window in London for £20 and found that it worked satisfactorily in a silver mine. After this success, engines were made not only for the mines but also for the Peruvian Mint. In 1816 Trevethick arrived in Peru and was given a tremendous reception by the populace including the ringing of bells; they even suggested putting up a statue of him in solid silver.

All went well until revolution broke out in South America under Simon Bolivar, the Liberator, but since Lima was the headquarters of the Spanish Army in South America at that time, calm remained in Peru until 1817 when the army sustained defeat by the rebels. To prevent the Spaniards from using the mines, the rebels threw all the machinery, including Trevethick's engines, down the shafts so that in 1824 the dismayed Trevethick was forced to leave Peru after seven years of toil. Undaunted, he went to Chile to investigate copper-mining and while there made £2500 by raising a sunken warship, only to lose it on worthless shares in a Panamanian pearl-fishing enterprise.

There then followed an amazing series of adventures. With a Scotsman called Gerard, Trevethick travelled to Ecuador on a mining scheme, then to Columbia at the request of Simon Bolivar. Finally they proposed to open up the mineral riches of Costa Rica hoping to construct railways, factories and plant so that 250 000 tons of ore a year would be produced. First they needed to return to England to raise capital and decided that, instead of making the long, time-consuming journey round Cape Horn, they would travel across the mountains and virgin forests to the Caribbean coast. After the most arduous journey in which there was loss of life and the near drowning of Trevethick, they arrived in rags and practically starved with no money for the voyage home.

Trevethick managed to get to the port of Cartagena and by chance met the young Robert Stephenson who gave him money for the voyage and bought him some decent clothes. When he reached England in 1827, he was penniless apart from some mining concessions in South America which no one seemed to be interested in. He was given a hero's welcome in Cornwall but that did not pay the bills. The remainder of his life was spent working on a variety of schemes including land-reclamation in Holland and a fantastic design for a gigantic tower to commemorate the passing of the Reform Bill in 1831. The column was to be 1000 ft high and 100 ft across the base, with a steam-engine-powered lift to take visitors to the top. It was to be made from 1200 pieces of cast-iron at a cost of £80 000, but the project was never realised.

In his last years living in a modest inn, Richard Trevethick worked in Dartford, Kent, with a young engineer called John Hall. He still designed machines until his death on 22 April 1833, when his workmen had to raise a subscription for his funeral so that he could be buried in a pauper's grave the location of which is now unknown.

Richard Trevethick was an impressive giant of a man over 6 ft 4 in tall with vivid blue eyes full of good humour and high spirits. He was a great optimist and had an unusual capacity for hard work. Towards the end of his life this truly great engineer wrote, "I have been branded with folly and madness for attempting what the world calls impossibilities." He ends, "However much I may be straitened in pecuniary circumstances, the great honour of being a useful subject can never be taken from me, which to me far exceeds riches."

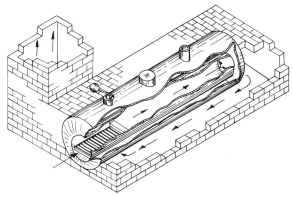

Trevethick's Cornish boiler

Giovanni Battista Venturi 1746-1822

Italian physicist

when 23 years old. He was later appointed Professor of Geometry and Philosophy at the University of Modena at the age of 27, ducal engineer and auditor and finally professor of physics. He devoted the greater part of his time to research into the flow of fluids and was a close associate of those giants of fluid mechanics, Daniel Bernoulli and Leonhard Euler.

Venturi experimented with the flow of water in pipes and open channels of tapering crossection and realised that a considerable amount of energy loss took place at sudden changes of crossection due to the formation of eddies. His study of the flow in such cases eventually led to the discovery of what is termed the 'Venturi Effect'; in addition he devised a number of very useful instruments for the determination of velocity, flow and pressure.

The 'venturi-meter' is a device well known to engineers which is widely used to measure the flow of liquids and gases. The so-called 'venturi-effect' is also used for example to measure the speed of aircraft, and the automobile engine carburrettor has an intake section called the 'venturi'.

These devices are all named after the famous Italian physicist Giovanni Battista Venturi who spent the greater part of his working life in Paris. Venturi was born in Northern Italy near Reggio on 11 September 1746, the son of Giovanni Domenico and Domenica (Gallian) Venturi. After receiving a parochial education he was ordained a priest

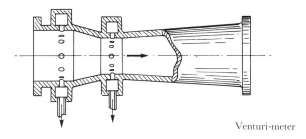

Venturi-meter

In a venturi, the flow velocity increases due to the reduced area with a corresponding decrease in pressure in the pipe as given by Bernoulli's equation. If a gradually tapering pipe is fitted downstream the pressure rises to its original magnitude apart from a small loss due to wall friction. Venturi discovered the effect but it was not until 1894, about 120 years later, that the hydraulics engineer Clemens Herschel invented the flowmeter which he named in honour of Venturi and which has been used extensively ever since by water authorities and in all fields of engineering.

Herschel (1842-1930) was educated at Harvard, Paris and Karlsruhe and having come under the influence of J. B. Francis turned to hydraulics at Lowell and then at Holyoke. He received the Elliott Cresson medal, like Pelton, for the invention described in his 1898 paper, 'The Venturi Water Meter', which was devised to measure the flow of water used by local power companies in pipes up to 2 ft in diameter. The name of the device originated from the use of the word 'Venturi' for the reading of the inlet/throat pressure difference. In the case of an open channel the device is known as a venturi flume and instead of a fall in pressure at the 'throat', there is a fall in the liquid level which may be measured to determine the rate of flow.

Little else is known of Venturi's activities apart from investigations which he carried out into the frequencies of audible tones, the movement of camphor on the surface of water, and the design of artillery. He died in Paris in 1822.

Count Alessandro Volta 1745-1827

Italian scientist

Volta's electric battery represents one of the most important discoveries and inventions of all time in the field of science. It is possibly more important than the telescope, microscope, steam engine or aeroplane, for from it stemmed the whole of electrical and electronic engineering. Electrical machines, electro-chemistry, illumination, electroplating, radio and telegraphy all owe their origins to the keen intellect of the scientist who first recognised this means of obtaining power electricity at a steady current.

Alessandro Volta was born on the shores of the beautiful Lake Como in Northern Italy on 18 February 1745, the son of a wealthy and aristocratic family. At school he showed a great enthusiasm for natural philosophy and learned French thoroughly so that he could study the tremendous scientific developments taking place in France at that time. Because he belonged to the aristocracy, Volta was able to devote a considerable amount of time to scientific experiments with the best of equipment. He was deeply interested in electricity and at the age of 24 wrote his first paper on that subject which he sent to the famous Italian scientist Beccaria.

As a result of this and other work, he soon acquired a name in the world of science and was appointed Professor of Physics at Como University at the age of 30. Three years later he took up a similar appointment at Pavia and was later made Rector. Volta's sympathies with Napoleon Boneparte, however, resulted in his dismissal from Pavia University in 1799 and he went to work in Paris where he met Napoleon. Apparently he made a good impression for in 1800, when Napoleon had conquered Northern Italy he ordered that Volta be reinstated as Rector.

Prior to Volta's experiments, the science of electricity had dealt only with the production of static electricity by friction machines in conjunction with Leyden Jar condensers, which produced very large voltages but only the minutest of currents. Galvani had shown in 1786 how the muscles of newly dissected frogs legs contracted in the presence of an electrical discharge; he also noticed that frogs legs hung on copper hooks had contracted when they touched an iron rail. Volta disagreed with Galvani's suggestion that the effect was due to the storage of electricity in the animal tissue and he later proved that electricity could be produced by dissimilar metals without resorting to the use of frog's legs.

In 1799, Alessandro Volta, now a Count, constructed his famous battery or 'Voltaic Pile', consisting of alternate discs of silver and zinc interspersed with absorbant material soaked with water. For the first time in history an amazingly large electrical current was available to man. In a letter to the President of the Royal Society in London, Sir Joseph Banks, Volta described his invention and thus gave to science a source of power which led to the immediate development of electrochemistry and eventually to electromagnetism and

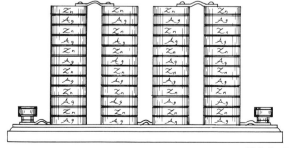

Voltaic pile

electric machines.

After his brilliant discovery Volta did little further work on his battery, leaving others to establish that it was the chemical reaction which produced the electricity. He retired from Pavia in 1815 and went home to Lake Como where he died on 5 March 1827.

The name of this great man is associated mainly with the SI and former unit of electric potential, the 'volt', from which a whole range of devices are named including the voltmeter, voltameter, voltaic pile and voltage divider.

James Watt 1736-1819

British craftsman and engineer

James Watt is regarded by many as the greatest of British Engineers, for not only did his invention of the improved steam engine lay the foundation of modern technology but it also led to mankind's control over the forces of nature to an extent previously undreamed of. His great success was due partly to his brilliant inventiveness and partly to his skill as a craftsman. His father's workmen were not exaggerating when they said of him, "Yon Jamie's gotten a fortune at his fingertips!."

James Watt was born in Greenock, a cradle of engineering and shipbuilding on the Renfrewshire coast in Scotland, on 19 January 1736. His father was a master shipwright and merchant and also a much respected Bailie and Treasurer of Greenock. James gained valuable experience in his father's workshop where he had his own workbench, with its kit of tools, and his own little forge. When he left grammar school, Watt worked for a time in Glasgow for an optician where he acquired skill as a maker of instruments. He then decided to go to London to make his fortune and set off on horseback on what was in those days a rather perilous twelve days' journey.

When James arrived in London he found himself without a job, as a friend of his father had promised to employ him but in the event let him down. It took a long time for the young man to find work but he was finally taken on by a maker of mathematical instruments who paid him a starvation wage. After a year of privation, James returned to Glasgow where he went into a partnership making instruments for the University. In 1746 he married his cousin, Margaret Millen.

One of Watt's first jobs was to design and build a perspective draughting machine, then in 1765 began the repair of a working model of a Newcomen steam engine. He hit upon the idea of improving the efficiency by fitting a steam jacket around the cylinder and of using a condenser for the spent steam to create a vacuum. In 1768, he met the Birmingham industrialist Matthew Boulton who had obtained a twenty-five year extension to the patent which Watt had taken out for his improved steam engine. The two men went into partnership to produce the new engine but it took many years to perfect and it was not until the 1780s that the firm made a profit.

From 1775 until 1790, Watt made many improvements to his engine. He also invented the double-acting engine, a rotative engine with sun and planet gear, an engine governor still today known as the Watt Governor, and the parallel link motion gear. He introduced the idea of measuring the power in terms of that of a horse by assuming that a mill horse walked in a 24 ft circle two and a half times a minute exerting a pull of 180 lb. The basis for these figures is unknown but they give a rate of working of 32 400 ft-lb/min which he later amended to 33 000. Another important invention was the steam engine indicator which showed exactly what was happening inside the cylinder while the engine was running.

Watt's first wife died in 1773 and, in 1776 he married Ann McGregor. While at Birmingham he took a prominent part in the famous Lunar Society where he met many eminent people including Priestly, Herschel and Wedgewood, the famous potter.

In the 1790s Watt and Boulton handed over much of their business to their sons and the firm became Boulton, Watt

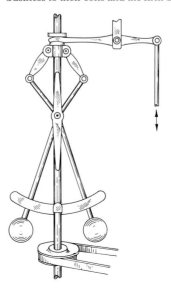

Watt's governor

and Sons. By 1800, Watt had practically retired but he continued inventing until his death. In his old age he produced an ingenious machine for copying sculptures. He died on 25 August 1819 at Heathfield, Birmingham.

Watt was fortunate to live to see the completion of his work and to receive the fame and monetary reward after a lifetime of struggle. His name will live forever, very appropriately in the unit of power, the watt, which replaces the horse-power originated by him. A monument was raised in Westminster Abbey with the inscription:

'To JAMES WATT who directing the force of an original genius early exercised in philosophic research to the improvement of the steam engine, enlarged the resource of his country, increased the power of man, and rose to an eminent place among the most illustrious followers of science, and the benefactors of the world.'

Wilhelm Eduard Weber 1804-1891

German scientist

Wilhelm Eduard Weber, founder of the modern system of electrical units, was born on 4 October 1804 at Wittenburg, Germany, the fifth child of a Professor of Divinity whose father had been a farmer. Wilhelm's brothers Ernst Heinrich and Eduard Friederich were both eminent anatomists. In his twenties, Wilhelm collaborated with Ernst on a treatise on wave motion, and with Eduard on a study of the mechanics of walking.

Wilhelm studied science at Halle where he eventually became a lecturer and then Assistant Professor. At the age of 27 he was called to the chair of Physics at the famous University of Göttingen at the suggestion of the great Karl Gauss with whom he later worked on studies of magnetism. In 1833, Weber's laboratory and Gauss' observatory were connected by Weber's electric telegraphy, the first ever practical system. It used only two wires for all the characters unlike other systems using one wire per character.

Weber was one of a group of academics who became known as the 'Göttingen Seven'. In 1837, Ernst August, King of Hanover, in which Göttingen is situated, abolished the parliamentary constitution by an autocratic decree in spite of his oath to uphold it. Seven members of the Göttingen staff including Weber and the brothers Grimm (of fairy-tale fame) drew up a manifesto denouncing the King's action and were immediately dismissed.

For five years Weber was without a post and a collection was made throughout the whole of Germany. However, he refused to live on the 1400 thalers collected and lived in near poverty. He carried on his researches with apparatus borrowed from Gauss. It was not until Weber was in his eighties that he used the money to buy apparatus after State economies had made it almost impossible for him to continue his work which included the establishment of an absolute unit of electric current which he based upon magnetic effects. In 1843 at the University of Leipzig, he devised a precision electrodynamometer based on the tangent galvanometer and made use of the voltameter, thus fixing the unit of current, the ampere.

After an absence of twelve years, Weber at last returned to Göttingen where he remained for the rest of his life; there he fixed the absolute unit of electromotive force, the volt, and was thus able to establish the absolute unit of resistance, the ohm. At about this time he developed the mirror-galvanometer which is of great value to this day, and a very important consequence of his work was the definition of the ratio of electrostatic to electromagnetic units of quantity known as Weber's Constant, which surprisingly turned out to be the velocity of light. This link between optics and electricity was later to be exploited by Maxwell in the discovery of radio waves.

Weber was by account a kind, friendly man, small in stature and very contented in spite of a lack of recognition. He was an enthusiastic walker right up to his eighties and was fervently patriotic, the unification of Germany in 1870 filling him with great joy. He never married and was looked after

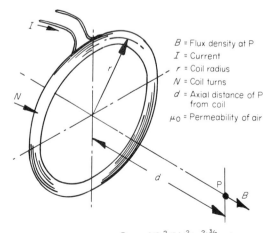

B = Flux density at P
I = Current
r = Coil radius
N = Coil turns
d = Axial distance of P from coil
μ_0 = Permeability of air

$B = \mu_0 NIr^2/2(d^2+r^2)^{3/2}$ webers

Flux density near coil

by a devoted niece until his death at the age of 86 at Göttingen on 23 June 1891.

In spite of his work on the unit of current, Weber was rather unfairly denied the honour of having his name used for it, the glory going to the late Ampère. In recent years, however, his name has been given to the SI unit of magnetic field strength, the weber (W), unfortunately at the expense of his great friend and patron Gauss whose name was used formerly.

Weber was undoubtedly the founder of precise measurement of electrical quantities such as current, emf, resistance and capacitance. His discovery of the connection between velocity of light and electricity combined the two laws of Coulomb for electrostatic and magnetic forces. He was also the first to define elementary electric particles and ascribe mass and charge to them.

Sir Joseph Whitworth 1803-1887

British engineer

Joseph Whitworth is known universally in connection with the establishment of a rationalised system of screw-threads and for the screw-thread which bears his name. It seems a great pity that, with the increasing use of metric screw-threads throughout the world, his illustrious name will disappear from numerous handbooks on engineering. The main contribution of this gifted engineer, however, was the introduction of new standards of accuracy in manufacturing to a degree hitherto undreamt of. In 1830 a good fitter could be expected to work to a sixteenth of an inch; ten years later, thanks to Joseph Whitworth, an accuracy of one ten-thousandth of an inch was a practical propostion.

Whitworth was born in the industrial town of Stockport, Cheshire on 21 December 1803, the son of Charles Whitworth a schoolmaster. After an undistinguished primary schooling, Joseph was bound apprentice to his uncle, a Derbyshire cotton spinner. After serving his four years apprenticeship Joseph worked for another four years as a mechanic in a Manchester factory before joining the famous engineer Henry Maudslay, the inventor of the screw-cutting lathe. In Maudslay's factory in Westminster Bridge Road, London, he worked with other men destined to be in the forefront of engineering achievement, such as James Nasmyth, inventor of the steam-hammer, and the brilliant inventor Richard Roberts.

Under Maudslay's wonderful tuition, Whitworth soon began to demonstrate his uncanny skill as a mechanic by constructing a truly plane surface and machine tools for turning, planing, milling, slotting, drilling, shaping and gear-cutting, all to a fantastic degree of accuracy; he also introduced the box casting for machine tool frames.

In 1841 Whitworth suggested the establishment of a uniform system of screw-threads with a fixed thread angle of 55° and a standard pitch for a given diameter. His work was displayed in the Great Exhibition of 1851 and his tools, micromeasuring techniques and screw-thread system earned him international fame, particularly his machine for measuring to a millionth of an inch.

During the 1850s, Whitworth turned his attention to the manufacture of guns, when he produced a novel form of rifled gun with a helical bore of hexagonal crossection which unfortunately suffered from fouling after repeated firing. There followed a considerable amount of bitter conflict with his rival, the great gunmaker Sir William Armstrong, over government contracts for armaments. This was finally solved when the two firms agreed to amalgamate in 1893 to form the great firm of Armstrong Whitworth.

Joseph Whitworth was a great champion of the cause of engineering education and to encourage talent in engineering he founded thirty scholarships costing £100 000 at what is now Manchester University and also at Manchester Technical School and Stockport Technical School. He wrote to Benjamin Disraeli in March 1868 concerning the scholarships saying that they were intended to, ". . . encourage students to combine practice with theory, and artisans to combine theory with perfection in workmanship." Unlike the attitude of many educators today, he was keen for all boys to learn the correct use of tools from the very beginning of their school careers.

After a lifetime of sterling service to British engineering, Sir Joseph Whitworth retired to Monte Carlo where he died at the age of 84 on 22 January 1887. During his lifetime Whitworth saw the development of modern manufacturing methods from the crude origins prevailing in the first quarter

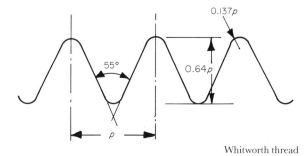

Whitworth thread

of the century. This humble mechanic had risen to become a Fellow of the Royal Society, a Baronet and the recipient of honorary degrees from Oxford University and Trinity College Dublin.

Thomas Young 1773-1829

English physicist, physician and linguist

Thomas Young qualified as a physician and linguist and took up physics almost as a hobby. He became well known throughout the world to engineers as the originator of Young's Modulus, but in the world of physics, however, he is famous for his discovery of the wave theory of light and his work on sound.

Thomas Young was born into a Quaker family of Milverton in Somerset, England, on 13 June 1773. He was a precocious child with a phenomenal memory and had mastered several languages at an extremely early age. He was attracted particularly to Middle Eastern languages and at fourteen had a good knowledge of Hebrew, Arabic and Syrian, as well as Greek and Latin, and by the age of seventeen he had read and absorbed Newton's *Principia* and *Optics*.

In spite of Young's predilection for physics, he was persuaded reluctantly by his great uncle to take up medicine and eventually take over his own lucrative practice in London. He entered St Bartholomew's Hospital and after only one year was made a Fellow of the Royal Society for a wonderful paper on the ciliary muscles of the eye. After leaving St Bartholomew's, Young studied at Edinburgh, Göttingen in Germany and finally at Cambridge, where he seemed to have plenty of time to devote to his experiments on light and sound. He practised as a physician for a time but when his uncle died and left him a fortune of £10 000 and a London house, he was able to concentrate entirely on his scientific experiments.

In his famous experiments on light in 1801, Young passed a beam of monochromatic light through two narrow slits close together and to his great delight observed light and dark bands on a screen. This proved conclusively his theory that light was transmitted in the form of waves and all that was needed was a simple calculation to determine the wavelength of any colour. In 1807 he carried out an investigation of the stress and strain in bars of metal subjected to loads. He found that for a given material there was a constant ratio between the load per unit area of a bar in tension and the extension per unit length due to the load. This is now called Young's Modulus and is of vital importance to the engineer.

The idea of the Modulus was introduced for the first time in his 'Course of Lectures on Natural Philosophy and the Mechanical Arts' in 1807. Young also investigated stresses due to impact loading and buckling of struts. He clearly demonstrated his practical engineering ability in work done for the Admiralty on the stiffening of ships' hulls by means of diagonal ribs. This great experimenter continued to publish papers on such diverse topics as, fluid mechanics, medicine and ancient languages. He was incidentally the first person to give the word 'energy' it's scientific significance.

Thomas Young's lifelong attraction for ancient languages led to his involvement in the attempts to decipher ancient Egyptian hieroglyphs and in 1814 he produced a fairly accurate translation of hieroglyphs and made a notable contribution to the reading of the Rosetta Stone.

Towards the end of his very full life, Young held a post as physician in St George's Hospital, London but retired in 1817 to carry out his duties as Commissioner for Weights and Measures and as the Foreign Secretary of the Royal Society. He died in London on 10 May 1829.

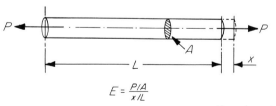

Young's modulus

Further Reading

AMPÈRE
ARAGO, D. F. J., *Oeuvres,* vols 1-3, Notices Biographiques (1854-62).
— Journal et Correspondance (1872).
DUNSHEATH, P. *A History of Electrical Engineering,* 57, Faber & Faber, London (1962).
GILLISPIE, C. C. (ed), *Dictionary of Scientific Biography,* **1,** 139-46, Chas. Scribner's Sons, New York (1971).
HART I. B., *Makers of Science,* OUP, London (1930).
LAUNAY, L. DE, *Le Grand Ampère,* Pessin, Paris (1925).
LENARD, P., *Great Men of Science,* 223-30, G. Bell & Sons, London (1933).
WILLIAMS, T. I. (ed), *A Biographical Dictionary of Scientists,* 10-11, A & C Black, London (1969).

ARCHIMEDES
BURSTALL, A. F. *A History of Mechanical Engineering,* 66-67, Faber & Faber, London (1963).
DONALDSON, A. C., 'Discharge tests on an Archimedean screw,' *Proc. I. Mech. E.,* **31,** 187-98 (1947).
DYKSTERHUIS, E. J. *Archimedes,* Munksgaard, Copenhagen (1956).
HEATH, T. L. *A History of Greek Mathematics,* OUP, Oxford (1921).
HEATH, T. L. *The Works of Archimedes with the Method of Archimedes,* Dover Publns Inc, New York.
HEATH, T. L. (ed), *The Works of Archimedes,* CUP (1897, supplement 1912).
MATSCHOSS, C. (trans Hatfield, H. S.) *Great Engineers,* 12-13, G Bell & Sons, London (1939).
ROUSE, H. AND INCE, S. *History of Hydraulics,* 15-22, 45, 48, 53, 79, Dover Publns Inc, New York (1963).
WILLIAMS, T. I. (ed) *A Biographical Dictionary of Scientists,* 16, A & C Black, London (1969).

ARMSTRONG
ARMSTRONG, W. G., 'Water pressure machinery,' *Proc I. Mech. E.,* 126 (1858), 21 (1868).
BURSTALL, A. F., *A History of Mechanical Engineering,* 297, 329, 330, Faber, London (1963).
DOUGAN, D., *The Great Gunmaker,* Frank Graham, Newcastle-upon-Tyne (1970).
EMERSON-TENNENT, SIR J., *The Story of Guns,* London 1864, *Elswick 1847-1947,* Vickers Brochure, Newcastle (1948).
GREAVES, W. F. AND CARPENTER, J. H., *A Short History of Mechanical Engineering,* 121, Longmans, London (1969).
SCOTT, J. D., *Vickers — A History,* Weidenfeld and Nicholson, London (1962).

BERNOULLI
DUGAS, R., *A History of Mechanics* (trans Maddox, J. R.), 233-4, 287-9. Editions du Griffon, Neuchatel. Switzerland (1955).
GILLISPIE, C. C. (ed), *Dictionary of Scientific Biography,* 36-46, Charles Scribner's Sons, New York (1971).
ROUSE, H. AND INCE, S., *History of Hydraulics,* Dover Publns. Inc, New York (1963).
SMITH, D. E. *History of Mathematics,* vol 1, Ginn & Co. (1923). Paperback edition, Dover Publns. Inc., New York (1958).
TIMOSHENKO, S. P. *History of Strength of Materials,* 27, 32, McGraw-Hill, New York (1953).
TOKATY, G. A., *A History & Philosophy of Fluid Mechanics,* 70, G. T. Foulis, Henley-on-Thames (1971).
WILLIAMS, T. I. (ed), *A Biographical Dictionary of Scientists,* 50-51, A & C Black, London (1969).

BESSEMER
BESSEMER, H., *An Autobiography,* Offices of 'Engineering', London (1905).
GREAVES, W. F. AND CARPENTER, J. H., *A Short History of Mechanical Engineering,* 13-14, 112-113, Longmans, London (1969).
MATSCHOSS, C. (trans Hatfield, H. S.), *Great Engineers,* 244-59, Bell & Sons, London (1939).
WILLIAMS, T. I. (ed), *A Biographical Dictionary of Scientists,* 57, A & C Black, London (1969).
WILLIAMS, T. I. (ed), *A History of Technology,* **5,** 53-7, Clarendon Press, Oxford (1978).

BOURDON
BOURDON, E., 'Description de manometres metalliques sans mercure, pour indiquer la pression de la vapeur dans les chaudieres,' *Bull. Soc. d'encouragement pour l'industrie nationale,* **50,** 197-200 (1851).
BRADSPIES, R. W., 'Bourdon Tubes', *Giannini Technical Notes,* Giannini Controls Corp., Duarte, Cali. (1961).
COWPER, C., 'On Bourdon's metallic barometer, indicator and other applications of the same principle,' *Proc. I. Mech. E.,* **3,** 141-150 (1852).
JENNINGS, F. B., 'Theories on Bourdon tubes,' *Trans. Amer. Soc. Mech. Engrs.,* **78,** 55-64 (1956).
OBITUARY, Eugene Bourdon. *Revue Scientifique,* 3rd ser., 21 (1884).
TRESCA, M. H., 'Notice sur M. Eugene Bourdon', *Bull. Soc. d'encouragement pour l'industrie nationale* (1884).

BRAMAH
BRADLEY, I., *A History of Machine Tools* (incl. Biography of J. Bramah), Model & Allied Pubns Ltd, Hemel Hempstead (1972).
BRAMAH, J., Patent No. 1177. 'A Water Closet upon a New Construction.' (1778).
DICKINSON, H. W., 'Joseph Bramah and his Inventions,' *Trans. Newcomen Soc.,* **22,** 169-86 (1941-2).
McNEIL, I., *Joseph Bramah, a Century of Invention,* 1749-1851, David & Charles, Newton Abbot (1968).
MATSCHOSS, C., *Great Engineers* (trans. Hatfield, H. S.), 192-4, G. Bell & Sons, London (1939).
SMILES, S., *Industrial Biography,* John Murray, London (1863). Reprinted David & Charles, Newton Abbot (1967).
TAYLOR, F. S., *British Inventions,* Longman (1950).
WILLIAMS, T. I. (ed), *A Biographical Dictionary of Scientists,* 79, A & C Black, London (1969).

CALLENDAR
CALLENDAR, H. L., 'Critical relations between water and steam,' *Proc. I. Mech. E.,* **117,** 811-38 (1929).
GILLISPIE, C. C. (ed), *Dictionary of Scientific Biography,* **3,** 14-20, Charles Scribner's Sons, New York (1971).
HOWARD, A. V. (ed), *Chambers' Dictionary of Scientists,* London (1950). *Dictionary of National Biography,* supplement 4, 152-4 (1930).
OBITUARY, Callendar, H. L., Mechanical Engineering, 129, *Amer. Soc. of Mech. Eng.* (1930).
OBITUARY, *Engineering,* **129,** 115-117 (1930).
OBITUARY, *Nature,* **125** 173-4 (1930).

CARNOT
CARNOT, N. L. S., *Reflections on the Motive Power of Fire,* with other papers by Clapeyron, E. and Clausius, R. Intro. by Mendoza, E., Dover Pubns Inc, New York (1960).
GILLISPIE, C. C. (ed), *Dictionary of Scientific Biography,* **3,** 70-84, Chas. Scribner's Sons, N.Y. (1971).
HOWARD, A. V. (ed), *Chambers Dictionary of Scientists,* London (1950).
LENARD, P., *Great Men of Science,* 231-5, G. Bell & Sons, London (1933).
MAGIE, W. F., *A Source Book in Physics,* 220, Harvard U.P., Cambridge, Mass. (1965).
MENDOZA, E., *Archives Internationales d'Histoire les Sciences,* **12,** 377 (1959).
WILLIAMS, T. I. (ed), *A Biographical Dictionary of Scientists,* 94-5, A & C Black, London (1969).

CASTIGLIANO
BURSTALL, A. F., *A History of Mechanical Engineering,* 290, Faber & Faber, London (1963).
CASTIGLIANO, C. A., *The Theory of Equilibrium of Elastic Systems and its Applications,* Dover Publns Inc, New York (1966).
CASTIGLIANO, C. A. (trans Andrews, E. S.), *Elastic Stresses in Structures,* Scott Greenwood, London (1919).
CASTIGLIANO, C. A. *Comptes rendus de l'Academie des Sciences,* **46,** 208 (1827).
CROTTI, F., Commemorazione di Alberto Castigliano, *Poltechnico,* 32, nos 11/12 (1884).
ORAWAS, G. Ae. AND McLEAN, L., 'Historical development of energetical principle in elastomechanics,' *Applied Mechanics Reviews* (1966).
TIMOSHENKO, S. P., *History of Strength of Materials,* 289-293, McGraw-Hill, New York (1953).

CELSIUS and FAHRENHEIT
ASIMOV, I., *Asimov's Biographical Encyc. of Science & Technology,* 159, Pan Books, London (1975).
GILLISPIE, C. C. (ed), *Dictionary of Scientific Biography,* **3,** 173-4, Charles Scribner's, New York (1971).
GRANDE LAROUSSE ENCYCLOPÉDIQUE, **2,** 741, Libraire Larousse, Paris (1960).
KNOWLES MIDDLETON, W. E., *A History of the Thermometer,* John Hopkins Press, Homewood, Baltimore (1967).
MOULTON, F. R. AND SCHIFFERES, J. J., *The Autobiography of Science,* 206, John Murray, London (1963).
NORDENMARK, N. V. E., *Anders Celsius,* Almqvist & Wiksells, Uppsala (1936).
WILLIAMS, T. I. (ed), *A Biographical Dictionary of Scientists,* 102, A & C Black, London (1969).
WOLF, A. *History of Science, Technology & Philosophy in the 18th Century,* Macmillan, London (1938).

BOYLE
FULTON, J. F., *A Bibliography of the Honourable Robert Boyle,* 2nd ed, Clarendon Press, Oxford (1961).
HALL, M. B., *Robert Boyle and Seventeenth Century Chemistry,* CUP, Cambridge (1958).
HART, I. B., *Makers of Science,* O.U.P., London (1930).
HUTCHINGS, D. (ed), *Late 17th Century Scientists,* Pergamon Press, London (1969).
MADDISON, R. E. W., *The Life of the Honourable R. Boyle,* FRS Taylor & Francis, London (1969).
MASSON, F., *Robert Boyle,* Constable, London (1914).
MORE, L. T., *Life and Works of the Honourable Robert Boyle,* OUP, Oxford (1944).
PILKINGTON, R., *Robert Boyle, Father of Chemistry,* John Murray, London (1959).
ROUSE, H. AND INCE, S., *History of Hydraulics,* 64, 65, 81, Dover Publns Inc, New York (1963).

CHARLES
ANNALES DE CHIMIE, **43,** 157 ff (1802).
ASIMOV, I., *Asimov's Biographical Encyclopaedia of Science & Technology,* **208,** Pan Books, London (1975).
FRANCE, ANATOLE, *L'Elvire de Lamartine. Notes sur M. et Mme Charles,* Paris (1893).
GILLISPIE, C. C. (ed), *Dictionary of Scientific Biography,* **3,** 207-8, Charles Scribner's Sons, New York (1971).
WILLIAMS, T. I. (ed), *A Biographical Dictionary of Scientists,* 105-6, A & C Black, London (1969).
WILLIAMS, T. I. (ed), *A History of Technology,* 395-6, Clarendon Press, Oxford (1978).

CLAPEYRON
BURSTALL, A. F., *A History of Mechanical Engineering,* 210-212, Faber & Faber, London (1963).
FOURCY A., *Histoire de l'École Polytechnique,* Paris (1828).
GILLISPIE, C. C. (ed), *Dictionary of Scientific Biography,* **3,** 286-7, Charles Scribner's Sons, N.Y. (1971).
KERKER, M., 'Sadi Carnot and the Steam Engineers,' *Isis,* **51,** 257-270 (1960).
TIMOSHENKO, S. P., *History of Strength of Materials,* 114-118, McGraw-Hill, New York (1953).

CLAUSIUS
BRUSH, S. G., 'The development of the kinetic theory of gases, 3 — Clausius,' *Annls. Science,* **14,** 185-96 (1958).
CLAUSIUS, R. J. E., *Abhandlungen über die mechanische Wärmetheorie* (1864-67). Trans W. R. Browne as *Mechanical Theory of Heat,* Macmillan, London (1879).
GILLISPIE, C. C., *Dictionary of Scientific Biography,* **3,** 303-10, Charles Scribner's Sons, New York (1971).
LENARD, P., *Great Men of Science,* 282, 296-8, G. Bell & Sons, London (1933).
MAGIE, W. F., *A Source Book in Physics,* 228, Harvard U.P., Cambridge, Mass. (1965).
PROC. ROYAL SOC. **48,** i (1890).
WILLIAMS, T. I. (ed), *A Biographical Dictionary of Scientists,* 108-9, A & C Black, London (1969).

CORIOLIS
ASIMOV, I., *Asimov's Biographical Encyclopaedia of Science & Technology,* 287-8, Pan Books, London (1975).
BURSTALL, A. F., *A History of Mechanical Engineering,* 253, Faber & Faber, London (1963).
DE CORIOLIS, G. G., 'Memoir sur les equations du mouvement relatif des systemes de corps,' *J. de l'Ecole Poly.,* **15,** 142-154 (1835).
DUGAS, R., *A History of Mechanics* (trans Maddox, J. R.), 374-9, Editions du Griffon, Neuchatel, Switz. (1955).
GILLISPIE, C. C. (ed), *Dictionary of Scientific Biography,* **3,** 416-9, Charles Scribner's Sons, New York (1971).
GRANDE LAROUSSE ENCYCLOPÉDIQUE, **3,** 509, Librairie Larousse, Paris (1960).
HALL, A. S., 'Teaching Coriolis' Law,' *J. Engng. Edn.,* **38,** 757-65 (1948).
ROUSE, H. AND INCE, S., *History of Hydraulics,* 150-1, 204, 241, Dover Publns. Inc., New York (1957).

COULOMB
COULOMB, C. A., Papers on Electricity and Magnetism (includes brief memoir), *Société Française de Physique,* Paris (1884).
DUNSHEATH, P., *A History of Electrical Engineering,* 291, Faber & Faber, London (1962).
GILLISPIE, C. C. (ed), *Dictionary of Scientific Biography,* **3,** 439-47, Charles Scribner's Sons, New York (1971).

GILLMOR, C. S., *Coulomb and the Evolution of Physics and Engineering in Eighteenth Century France,* Princeton U.P., Princeton (1971).
HEYMAN, J., *Coulomb's Memoirs on Statics,* CUP, Cambridge (1972).
LENARD, P., *Great Men of Science,* 100, 149-158, G. Bell & Sons, London (1933).
TIMOSHENKO, S. P., *History of Strength of Materials,* 47-54, McGraw-Hill, New York (1953).
WILLIAMS, T. I., *A Biographical Dictionary of Scientists,* 117, A & C Black, London (1969).

CURTIS

EMMET, W. L. R., 'Curtis steam turbine,' *Proc. I. Mech. E.,* **67,** 715-735 (1904).
ROBINSON, E. L., 'The steam turbine in the United States, 3 — Developments by the General Electric Company,' *Mech. Engng.* **59,** 239-56 (1937).

D'ALEMBERT

DUGAS, R., D'Alembert et l'essai d'une nouvelle theorie sur la resistance des fluides,' *Bull. Société Française des Mecaniciens,* **2,** No. 6 (1952).
GILLISPIE, C. C. (ed), *Dictionary of Scientific Biography,* **1,** 110-17, Chas. Scribner's Sons, New York (1971).
HANKINS, T. L., *Jean D'Alembert, Science & the Enlightenment,* Clarendon Press, Oxford (1970).
LENARD, P., *Great Men of Science,* 96, G. Bell & Sons, London (1933).
ROUSE, H. AND INCE, S., *History of Hydraulics,* 94-106, Dover Pubns. Inc., New York (1957).
SMITH, D. E., *History of Mathematics,* vol. 1, Ginn & Co, London (1923). Paperback edition, Dover Publications, London (1958).
WILLIAMS, T. I. (ed), *A Biographical Dictionary of Scientists,* 127, A & C Black, London (1969).

DARCY and FANNING

BURSTALL, A. F., *A History of Mechanical Engineering,* 322, Faber & Faber, London (1963).
DARCY, H. P. G., 'Recherches experimentales relatives au mouvement de l'eau dans les tuyaux,' Paris (1857).
ENCYCLOPAEDIA BRITANNICA, Micropaedia, **3,** 377.
ROUSE, H. AND INCE, S., *History of Hydraulics,* 169-208, Dover Pubns. Inc., New York (1963).
TOKATY, G. A., *A History & Philosophy of Fluid Mechanics,* 93, 97, G. T. Foulis & Co, Henley (1971).
WILLIAMS, T. I. (ed), *A History of Technology,* **5,** 539, 546, 563, Clarendon Press, Oxford (1978).

DE LAVAL

BURSTALL, A. F., *A History of Mechanical Engineering,* 339-40, Faber & Faber, London (1963).
CENTENNIAL MEMORIAL PUBLICATION, The Laval Separator Co., Stockholm (1915).
GREAVES, W. F. AND CARPENTER, J. H., *A Short History of Mechanical Engineering,* 73-4, Longmans, London (1969).
KENNEDY, R. C. E., *Modern Engines and Power Generators,* Vol. 1, Caxton Publishing Co., London (c. 1905).
LEA, E. S. AND MEDEN, E., 'De Laval steam turbine,' *Proc. I. Mech. E.,* **67,** 697-714 (1904).
SMITH, G. W., Dr Carl Gustav Patrik de Laval (1845-1913) and the three de Laval companies in North America, Newcomen Soc. N. Amer., New York (1954).
WILLIAMS, T. I. (ed), *A Biographical Dictionary of Scientists,* 139, A & C Black, London (1969).
WILLIAMS, T. I. (ed), *A History of Technology,* **5,** 138, Clarendon Press, Oxford (1978).

DESCARTES

BECK, L. J., *The Method of Descartes,* Clarendon Press, Oxford (1952).
HARNE, R. (ed), *Early 17th Century Scientists,* 159, Pergamon, London (1965).
LENARD, P., *Great Men of Science,* 51-53, G. Bell & Sons, London (1933).
ROUSE, H. & INCE, S., *History of Hydraulics,* 45, 74-89, 108, 123, 202, Dover Publns. Inc., New York (1957).
SCOTT, J. F., *The Scientific Works of René Descartes,* Taylor & Francis, London (1976).
SMITH, D. E., *History of Mathematics,* 2 vols, 371, Dover Pubns. Inc., New York (1951).
WILLIAMS, T. I. (ed), *A Biographical Dictionary of Scientists,* 140-142, A & C Black, London (1969).

DIESEL

ASIMOV, I., *Asimov's Biographical Encyclopaedia of Science & Technology,* 500-1, Pan Books, London (1975).
DIESEL, R., *Theory and Construction of a Rational Heat Motor,* trans. B. Donkin, Syon, London (1894).
DONKIN, B., *Gas, Oil and Air Engines,* Charles Griffin and Co, London (1911).
GOLDBECK, G. AND SCHILDBERGER, F., *From Engines to Autos,* Henry Regnery, Chicago (1960).
GREAVES, W. F. AND CARPENTER, J. H., *A Short History of Mechanical Engineering,* 85, Longmans, London (1969).
GROSSER, M., *Diesel: the Man and the Engine,* David & Charles, Newton Abbot (1980).
MATSCHOSS, C., *Great Engineers,* 299-304, Bell & Sons, London (1939).
MOON, J. F., *Rudolf Diesel and the Economic Power Unit* (Pioneers of Science and Discovery), Priory Press.
NITSKE, W. R. AND WILSON, C. M., *Rudulph Diesel: Pioneer of the Age of Power,* Univ. of Oklahoma Press, Norman, Oklahoma (1965).
WILLIAMS, T. I. (ed), *A Biographical Dictionary of Scientists,* 146, A & C Black, London (1969).
WILLIAMS, T. I. (ed), *A History of Technology,* **5,** 163, **7,** 141, Clarendon Press, Oxford (1978).

ERICSSON

ASIMOV, I., *Asimov's Biographical Encyclopaedia of Science & Technology,* 313, Pan Books, London (1975).
BURSTALL, A. F., *A History of Mechanical Engineering,* 273, 277, 288, 350, Faber & Faber, London (1963).
CHURCH, W. C., *Life of John Ericsson,* 2 vols, Sampson Lowe & Marston, London (1892).
FINKELSTEIN, T., 'Air Engines,' *The Engineer,* **207,** 492-6 (1959).
GREAVES, W. F. AND CARPENTER, J. H., *A Short History of Mechanical Engineering,* 80, Longmans, London (1969).
MATSCHOSS, C., *Great Engineers,* 204-216, Bell & Sons, London (1939).
ROUSE, H. AND INCE, S., *History of Hydraulics,* 155-6, Dover Publns. Inc., New York (1963).
WILLIAMS, T. I. (ed), *A Biographical Dictionary of Scientists,* 167, A & C Black, London (1969).

EULER

DUGAS, R., *A History of Mechanics* (trans Maddox, J. R.), Editions du Griffon, Neuchatel, Switz. (1955).
GILLISPIE, C. C. (ed), *Dictionary of Scientific Biography,* **4,** 467-84, Charles Scribner's Sons, New York (1971).
ROUSE, H. AND INCE, S., *History of Hydraulics,* Dover Publns. Inc., New York (1963).
SPIESS, O., *Leonhard Euler,* Huber, Frauenfeld (1929).
TIMOSHENKO, S. P., *History of Strength of Materials,* 28; 36, McGraw-Hill, New York (1953).

TURNBULL, H. W., *The Great Mathematicians,* 107, Methuen, London (1929).
WILLIAMS, T. I. (ed), *A Biographical Dictionary of Scientists,* 169-70, A & C Black, London (1969).

FARADAY
AGASSI, J., *Faraday as a Natural Philosopher,* Univ. of Chicago Press, Chicago and London (1971).
BOWERS, B., *Michael Faraday & Electricity* (Pioneers of Science & Discovery), Priory Press, London (1974).
KENDALL, J., *Michael Faraday,* Faber & Faber, London (1957).
LENARD, P., *Great Men of Science,* 247-63, Bell & Sons, London (1933).
WILLIAMS, L. P., *Michael Faraday, a Biography,* Chapman & Hall, London (1965).
WILLIAMS, L. P., *The Origins of Field Theory,* Chapman & Hall, London (1966).
WILLIAMS, L. P., *The Selected Correspondence of Michael Faraday,* 2 vols, CUP (1971).
WILLIAMS, T. I. (ed), *A Biographical Dictionary of Scientists,* 174-6, A & C Black, London (1969).

FOURIER
ARAGO, F., *Joseph Fourier,* Smithsonian Annual Report, 136 (1871).
CHAMPOLLION-FIGEAC, J. J., *Fourier, Napoleon, l'Egypte et les Cent Jours,* Paris (1844).
DARBOUX, J. G. (ed), *J. Fourier, Oeuvres,* 2 vols, Gauthier-Villars, Paris (1888-90).
HERIVEL, J., *Joseph Fourier, The Man and the Physicist,* Clarendon Press, Oxford (1975).
LENARD, P., *Great Men of Science,* 234, Bell & Sons, London (1933).
MAUGER, G., *Joseph Fourier,* Annuaire Statistique de l'Yonne, 270-76 (1837).
ROUSE, H. AND INCE, S., *History of Hydraulics,* 213, Dover Publns. Inc., New York (1963).
WILLIAMS, T. I. (ed), *A Biographical Dictionary of Scientists,* 190, A & C Black, London (1969).

FOURNEYRON
BURSTALL, A. F., *A History of Mechanical Engineering,* 248, Faber & Faber, London (1963).
CROZET-FOURNEYRON, M., *Invention de la turbine,* Beranger, Paris (1924).
GILLISPIE, C. C. (ed), *Dictionary of Scientific Biography,* **5,** 101, Charles Scribner's Sons, New York (1971).
GREAVES, W. F. AND CARPENTER, J. H., *A Short History of Mechanical Engineering,* 53, Longmans, London (1969).
KEATUR, F. W., 'Benoit Fourneyron, 1802-1867.' *Mechanical Engineering,* **61,** 295-301 (1939).
ROUSE, H. AND INCE, S., *History of Hydraulics,* 146-7, 164, 165, Dover Publns. Inc., New York (1963).
WILLIAMS, T. I. (ed), *A Biographical Dictionary of Scientists,* 190-1, A & C Black, London (1969).

FRANCIS
BARNA, P. S., *Fluid Mechanics for Engineers,* 358, Butterworths, London (1957).
BURSTALL, A. F., *A History of Mechanical Engineering,* 243, 247, 280, Faber & Faber, London (1963).
ENCYCLOPAEDIA BRITANNICA, Micropaedia, **4.**
ENCYCLOPAEDIA AMERICANA, **2..**
FINNIECOMBE, J. R., 'Development of the Kaplan turbine,' *Engineering,* **150,** 381-383 (1940). **152,** 21-24 (1941).
FRANCIS, J. B., *Lowell Hydraulics Experiments,* New York (1855).
FRISCH, J., 'The Kaplan turbine, design and trends, *Trans. Amer. Soc. Mech. Engrs.,* **76,** 765-73 (1954).
McCORMACK, W. J., 'Performance characteristics of Francis type turbines,' *Am. Soc. Mech. Engrs.,* **78,** 417-26 (1956)
PRAZIL, F., 'Results of Experiments with Francis turbines and tangential (Pelton) turbines,' *Proc. I. Mech. E.,* **81,** 647-66 (1911).
ROUSE, H. AND INCE, S., *History of Hydraulics,* 165, 166, 189, Dover Publns. Inc., New York (1963).

FROUDE
ABELL, W., 'William Froude,' *Trans. Inst. Naval Arch.,* **76** (1934).
BARNA, P. S., *Fluid Mechanics for Engineers,* 146, Butterworths, London (1957).
BURSTALL, A. F., *A History of Mechanical Engineering,* 324, 360, Faber & Faber, London (1963).
FROUDE, W., 'Experiments on the surface friction experienced by a plane moving through water, *Brit. Ass. Adv. Sc.,* 42nd meeting (1872).
FROUDE, W., 'On experiments with H.M.S. *Greyhound,*' *Trans. Inst. Nav. Arch.,* **16** (1874).
FROUDE, W., 'Dynamometer for large screw steamers,' *Proc. I. Mech. E.,* **28,** 237-75 (1877).
'MEMOIRS OF THE LATE WILLIAM FROUDE, LLD, FRS,' *Trans. Inst. Naval. Arch.,* **20,** 264-269 (1879).
ROUSE, H. AND INCE, S., *History of Hydraulics,* 155-6, 182-87, 191, 236, Dover Publns, Inc., New York (1963).
'THE FROUDE HYDRAULIC DYNAMOMETER,' *Edgar Allen News,* **22,** 193-7 (1943).
THE LIFE AND WORK OF WILLIAM FROUDE, Heenan & Froude Ltd, Worcester.

GALVANI
DUNSHEATH, P., *A History of Electrical Engineering,* 28, Faber & Faber, London (1962).
GILLISPIE, C. C. (ed), *Dictionary of Scientific Biography,* **5,** 267-8, Charles Scribner's Sons, New York (1971).
LENARD, P., *Great Men of Science,* 158-63, Bell & Sons, London (1933).
MAGIE, W. F., *A Source Book in Physics,* 420, Harvard Univ. Press, Cambridge, Mass. (1965).
MAJORANA, W., 'Commemorazione di Luigi Galvani,' *Nuovo Cimento,* **14** (1937).
WILLIAMS, T. I. (ed), *A Biographical Dictionary of Scientists,* 203-4, A & C Black, London (1969).

GAUSS
BELL, E. T., *Men of Mathematics,* vol. 1 (1937).
DUNNINGTON, C. W., *K. F. Gauss, Titan of Science,* Exposition Press, New York (1955).
GILLISPIE, C. C. (ed), *Dictionary of Scientific Biography,* **5,** 298-315, Charles Scribner's Sons, New York (1971).
HALL, T., *Karl Friedrich Gauss: a Biography,* MIT Press, Cambridge, Mass. (1970).
LENARD, P., *Great Men of Science,* 240-7, Bell & Sons, London (1933).
MAGIE, W. F., *A Source Book of Physics,* Harvard U.P., Cambridge, Mass., 519 (1965).
SMITH, D. E., *History of Mathematics,* Dover Publns. Inc., 502 (1951).
WILLIAMS, T. I. (ed), *A Biographical Dictionary of Scientists,* 207-9, A & C Black, London (1969).

HENRY
A MEMORIAL OF JOSEPH HENRY, Govt. Printing Office, Washington (1880).
ASIMOV, I., *Asimov's Biographical Encyclopaedia of Science and Technology,* 287-8, Pan Books, London (1975).
COULSON, T., *Joseph Henry: his Life and Work,* Princeton Univ. Press, Princeton (1950).

DUNSHEATH, P., *A History of Electrical Engineering,* 62, Faber & Faber, London (1962).
GILLISPIE, C. C. (ed), *Dictionary of Scientific Biography,* **6,** 227-81, Charles Scribner's Sons, New York (1971).
REINGOLD, N. (ed), *Science in Nineteenth Century America,* Macmillan, London (1966).
THE SCIENTIFIC WRITINGS OF JOSEPH HENRY, Smithsonian Institute Publication (1886).
WILLIAMS, T. I. (ed), *A Biographical Dictionary of Scientists,* 250-1, A & C Black, London (1969).

HERO OF ALEXANDRIA

BRADY, J. F., 'The rotating nozzle (Hero's) turbine,' *Am. Soc. Engrs.,* Paper 60-WA-293 (1960).
BURSTALL, A. F., *A History of Mechanical Engineering,* Faber & Faber, London (1963).
DRACHMANN, A. G., *Ktesibios, Philon and Heron,* Munksgaard, Copenhagen (1948).
HEATH, T. L., *A History of Greek Mathematics,* vol. 12 (1921).
LANDELS, J. C., *Engineering in the Ancient World,* Chatto & Windus, London (1978).
MATSCHOSS, C., *Great Engineers,* 16-21, Bell & Sons, London (1939).
ROUSE, H. AND INCE, S., *History of Hydraulics,* 19-22, 32, 45, 48, 52, 106, Dover Publns. Inc., New York (1963).
SINGER, C., HOLMYARD, E. J., HALL, A. R. AND WILLIAMS, T. I. (ed), *A History of Technology,* **2,** OUP (1956).
WILLIAMS, T. I. (ed), *A Biographical Dictionary of Scientists,* 251-2, A & C Black, London (1969).

HERTZ

APPLEYARD, R., *Pioneers of Electrical Communication,* London (1930).
BURSTALL, A. F., *A History of Mechanical Engineering,* 290, Faber & Faber, London (1963).
ENDEAVOUR, **16,** 3 (1957).
LENARD, P., *Great Men of Science,* 358-71, Bell & Sons, London (1933).
TIMOSHENKO, S. P., *History of the Strength of Materials,* 347-53, McGraw-Hill, New York (1953).
WILLIAMS, T. I. (ed), *A Biographical Dictionary of Scientists,* 254-5, A & C Black, London (1969).

HOOKE

BURSTALL, A. F., *A History of Mechanical Engineering,* 144-167, 255, Faber & Faber, London (1963).
ESPINASSE, M., *Robert Hooke,* Heinemann, London (1956).
GUNTER, R. T., *Early Science in Oxford,* vols 6, 7: *The Life and Work of Robert Hooke,* the author, Oxford (1930). Reprinted by Dawson, London (1968).
KEYNES, G. A., *A Bibliography of Dr Robert Hooke,* Clarendon Press, Oxford (1960).
LENARD, P., *Great Men of Science,* 72, 109, 123, Bell & Sons, London (1933).
ROBINSON, H. W. AND ADAMS, W. (eds), *The Diary of Robert Hooke,* Wykeham Pubns, London (1968).
ROUSE, H. AND INCE, S., *History of Hydraulics,* 81, 83, Dover Publns. Inc., New York (1963).
TIMOSHENKO, S. P., *History of the Strength of Materials,* 17-20, McGraw-Hill, New York (1953).
WILLIAMS, T. I. (ed), *A Biographical Dictionary of Scientists,* 264-5, A & C Black, London (1969).

JOULE

BURSTALL, A. F., *A History of Mechanical Engineering,* 287, Faber & Faber, London (1963).
GILLISPIE, C. C. (ed), *Dictionary of Scientific Biography,* **7,** 180-2, Charles Scribner's Sons, New York (1971).
JOULE, J. P., *Scientific Papers,* 2 vols, London (1884-7).

LENARD, P., *Great Men of Science,* 286-92, Bell & Sons, London (1933).
MAGIE, W. F., *A Source Book of Physics,* 172, 203, 255, 524, Harvard U.P., Cambridge, Mass. (1965).
NORTH, J. (ed), Mid 19th Century Scientists Pergamon Press. London (1969).
REYNOLDS, O., *Memoir of James Prescott Joule,* Manchester (1892).
STEFFENS, H. S., *James Prescott Joule and the Concept of Energy,* Dawsons, Folkestone (1979).
THE SCIENTIFIC PAPERS OF JAMES PRESCOTT JOULE, The Physics Society, London (1884).
WILLIAMS, T. I. (ed), *A Biographical Dictionary of Scientists,* 281-2, A & C Black, London (1969).

KÁRMÁN

BARNA, P. S., *Fluid Mechanics for Engineers,* 51, 166, 201, Butterworths, London (1957).
BURSTALL, A. F., *A History of Mechanical Engineering,* 398 (1963), Faber & Faber, London (1963).
GOLDSTEIN, S. (ed), *Modern Developments in Fluid Mechanics,* OUP, Oxford (1938).
KÁRMÁN, T. VON AND BIOT, M. A., *Mathematical Methods in Engineering,* McGraw-Hill, New York (1940).
KÁRMÁN, T. VON., *Aerodynamics,* Cornell Univ. Press, Ithica, New York (1954).
KÁRMÁN, T. VON AND EDSON, L., *The Wind and Beyond: Theodore von Kármán, Pioneer in Aviation and Pathfinder in Space,* Little, Brown, Boston (1967).
ROUSE, H. AND INCE, S., *History of Hydraulics,* 233-242, Dover Publns. Inc., New York (1963).

KELVIN

BURSTALL, A. F., *A History of Mechanical Engineering,* 278, 287, 312, 353, 443, Faber & Faber, London (1963).
GRAY, A., *Lord Kelvin,* London (1908). Reprinted Chelsea Publing Co, New York (1973).
LENARD, P., *Great Men of Science,* 298-307, Bell & Sons, London (1933).
Nature, **77,** 157 (1907).
Proc. Roy. Soc., **81 A,** iii (1908).
ROUSE, H. AND INCE S., *History of Hydraulics,* 211, 212, Dover Publns. Ince., New York (1963).
RUSSELL, A., *Lord Kelvin,* Blackie, London (1938).
THOMPSON, S. P., *Life of Lord Kelvin,* 2 vols, Macmillan, London (1910).

KIRCHHOFF

GILLISPIE, C. C. (ed), *Dictionary of Scientific Biography,* **7,** 379-83, Charles Scribner's Sons, New York (1971).
LENARD, P., *Great Men of Science,* 324-338, Bell & Sons, London (1933).
Proc. Roy. Soc., **46,** vi (1889).
ROUSE, H. AND INCE, S., *History of Hydraulics,* 200-1, 205, 211, Dover Pubns. Inc., New York, 1963.
TIMOSHENKO, S. P., *History of Strength of Materials,* 252-55, McGraw-Hill, New York (1953).
WILLIAMS, T. I. (ed), *A Biographical Dictionary of Scientists,* 290-1, A & C Black, London (1969).

LAGRANGE

BALL, W. W. R., *A Short Account of the History of Mathematics,* Dover Pubns Inc., New York (1908).
GILLISPIE, C. C. (ed), *Dictionary of Scientific Biography,* **7,** 559-73, Charles Scribner's Sons, New York (1971).
Oeuvres Complètes (inc. biography by J. B. J. Delambre), Gauthier-Villars, Paris (1867-92).

ROUSE, H. AND INCE, S., *History of Hydraulics,* 18, 107-11, 195, 199, 200, 211, Dover Pubns. Inc., New York (1963).
STRUICK, D. J., *Concise History of Mathematics,* Dover Pubns. Inc., New York (1954).
TIMOSHENKO, S. P., *History of the Strength of Materials,* 37-40, McGraw-Hill, New York (1953).
TURNBULL, H. W., *The Great Mathematicians,* p. 114, Methuen, London (1929).
WILLIAMS, T. I. (ed), *A Biographical Dictionary of Scientists,* 302-3, A & C Black, London (1969).

LAMÉ

BURSTALL, A. F., *A History of Mechanical Engineering,* 210, Faber & Faber, London (1963).
Catalogue of Scientific Papers, Royal Society (Bibliography) III, VIII and X.
GILLISPIE, C. C. (ed), *Dictionary of Scientific Biography,* **7,** 601-2, Charles Scribner's Sons, New York (1971).
POGGENDORF, I., pt. 1, 1359-60.
TIMOSHENKO, S. P., *History of Strength of Materials,* 114-18, McGraw-Hill, New York (1953).

LAPLACE

BALL, W. W. R., *A Short Account of the History of Mathematics,* Dover Pubns. Inc. (1908).
CROSLAND, M., *The Society of Arcueil,* Heinemann, London (1967).
DOIG, P., *A Concise History of Astronomy,* Chapman & Hall, London (1950).
GILLISPIE, C. C. (ed), *Dictionary of Scientific Biography,* **7,** 30, Charles Scribner's Sons, New York (1971).
LENARD, P., *Great Men of Science,* 218-223, Bell & Sons, London (1933).
PANNAKOCK, A., *A History of Astronomy,* Allen & Unwin, London (1961).
ROUSE, H. AND INCE, S., *History of Hydraulics,* 108-9, 142, 195, 199, 211, Dover Pubns. Inc., New York (1963).
WHITTAKER, E. T., *American Mathematical Monthly,* **56,** 369 (1949).
WILLIAMS, T. I. (ed), *A Biographical Dictionary of Scientists,* 308-10, A & C Black, London (1969).

LENOIR

BURSTALL, A. F., *A History of Mechanical Engineering,* 333, Faber & Faber, London (1963).
DONKIN, B., *Gas, Oil and Air Engines,* Charles Griffin & Co, London (1911).
EVANS, A. F., *The History of the Oil Engine.*
GREAVES, W. F. AND CARPENTER, J. A., *A Short History of Mechanical Engineering,* 80, Longmans, London (1969).
KASTNER, L. J., Century in the History of Reciprocating Internal Combustion Engine, *Proc. I.Mech.E.,* **169,** 303 (1955).
WILLIAMS, T. I. (ed), *A Biographical Dictionary of Scientists,* 322-23, A & C Black, London (1969).

LEONARDO DA VINCI

GIBBS-SMITH, C., *The Inventions of Leonardo da Vinci,* Phaedon, Oxford (1978).
HART, I. B., *The Mechanical Investigations of Leonardo da Vinci,* Chapman & Hall, London (1925).
LENARD, P., *Great Men of Science,* 9-12, Bell & Sons, London (1933).
MacCURDY, E., *The Notebooks of Leonardo da Vinci,* 2 vols, Reprint, Society, London (1938).
MATSCHOSS, C., *Great Engineers.* 48-60, Bell & Sons, London (1939).
McMURRICH, J. P., *Leonardi d Vinci the Anatomist,* Williams & Wilkins, Baltimore (1930).
RETI, L., *The Unknown Leonardo,* McGraw-Hill (1974).
ROUSE, H. & INCE, S., *History of Hydraulics,* 42-52, 57-60, 67, 69, 79, 135, 137, Dover Pubns. Inc. (1963).

SMITH, D. E., *History of Mathematics,* 2 vols, 254, 294, Dover Publns. Inc., New York (1951).
TIMOSHENKO, S. P., *History of Strength of Materials,* 2-6, McGraw-Hill, New York (1953).
TOKATY, G. A., *A History & Philosophy of Fluid Mechanics,* 36, 42, G. T. Foulis, Henley (1971).
WILLIAMS, T. I. (ed), *A Biographical Dictionary of Scientists,* 323-24, A & C Black, London (1969).

MACH

BLACKMORE, J. T., *Ernst Mach, his Work, Life and Influence,* Univ. of California Press, Berkeley (1972).
DUGAS, R. A., *A History of Mechanics,* 443-4, Editions du Griffon, Neuchatel, Switzerland (1955).
FRANK, P. G., 'Einstein, Mach and logical positivism'. In Schlipp, P. A. (ed), *Albert Einstein: Philosopher-Scientist* (1959).
Grande Larousse Encyclopédique, **6,** Libraire Larousse, Paris.
GILLISPIE, C. C. (ed), *Dictionary of Scientific Biography,* **8,** 595-607, Charles Scribner's Sons, New York (1971).
MACH, E., *The Science of Mechanics,* The Open Court Pubg. Co., La Salle, Illinois (1960).
ROUSE, H. AND INCE, S., *History of Hydraulics,* 195, Dover Publns. Inc., New York (1963).
TOKATY, G. A., *A History and Philosophy of Fluid Mechanics,* 193, G. T. Foulis & Co., Henley (1971).
WILLIAMS, T. I. (ed), *A Biographical Dictionary of Scientists,* 345-46, A & C Black, London (1969).

MAGNUS

BARNA, P. S., *Fluid Mechanics for Engineers,* 217, Butterworths, London (1957).
BURSTALL, A. F., *A History of Mechanical Engineering,* 422, Faber & Faber, London (1963).
OBITUARY, H. G. MAGNUS, *Engineering* (1870).
SWANSON, W. M., 'The Magnus Effect: A Summary of Investigations to date,' *Basic Engineering,* **83,** *Trans. A.S.M.E., series D,* 461-70 (1966).
TOKATY, G. A., *A History of Philosophy of Fluid Mechanics,* 141, 147, 151, G. T. Foulis & Co., Henley (1971).

MAXWELL

BURSTALL, A. F., *A History of Mechanical Engineering,* 288, 315, 375, Faber & Faber, London (1963).
CAMPBELL, L. AND GARNETT, W., *Life of Maxwell* (1882), reprinted Johnson Reprint, New York (1969).
DOMB, C., *Clerk Maxwell and Modern Science,* Athlone Press, London (1963).
LENARD, P., *Great Men of Science,* 339-43, G. Bell & Sons, London (1933).
MAY, C. P., *James Clerk Maxwell & Electromagnetism,* Chatto & Windus, London (1964).
NIVEN, W. D. (ed), *The Scientific Papers of James Clerk Maxwell,* 2 vols, Dover Pubns. Inc., New York (1965).
Proc. Roy. Soc., **33,** i (1882).
SMITH, D. E., *History of Mathematics,* 2 vols, 470, Dover Pubns. Inc., New York (1951).
TIMOSHENKO, S. P., *History of Strength of Materials,* 268-275, McGraw-Hill, New York (1953).
WILLIAMS, T. I., *A Biographical Dictionary of Scientists,* 357-8, A & C Black, London (1969).

MOHR

BURSTALL, A. F., *A History of Mechanical Engineering,* 290, Faber & Faber, London (1963).
GEHLER, W., Christian Otto Mohr; 3 articles in *Festschrift Otto Mohr zum 80 Geburtstag,* Berlin (1916).

GILLISPIE, C. C. (ed), *Dictionary of Scientific Biography,* **9,** 445-6, Charles Scribner's Sons, New York (1971).
OTTO MOHR, *Zeitschrift des Vereins deutscher Ingenieure,* **62,** 114 (1918).
TIMOSHENKO, S. P., *History of Strength of Materials,* 283-8, McGraw-Hill, New York (1953).
TODHUNTER, I. AND PEARSON, K., *History of the Theory of Elasticity,* 2 vols, CUP, Cambridge (1886 and 1893).

MOLLIER

Brockhaus Enzyklopädie, **12,** 717, F. A. Brockhaus, Wiesbaden (1971).
Encyclopaedia Britannica, Micropaedia, **6,** 981.
ELSNER, N., Richard Mollier des Mensch und Wissenshafter Zeitschrift der Technischen, Universität, Dresden, **13,** 1101-1103 (1964).
GILLISPIE, C. C. (ed), *Dictionary of Scientific Biography,* **9,** 46, Charles Scribner's Sons, New York (1971).
MOLLIER, R., *Neue Tabellen und Diagramme für Wasserdampf,* 7th edn, Springer, Berlin (1932).
PLANK, R., 'Richard Mollier,' *Zeit. für die gesamte Kalteindustrie,* 717, **15,** (1963).

NEWTON

ANDRADE, E. N. DA C., *Sir Isaac Newton,* Collins, London (1954).
EDLESTON, J., *Correspondence of Sir Isaac Newton and Professor Coates,* Frank Cass & Co, London (1969).
HERIVEL, J., *The Background to Newton's 'Principia',* Clarendon Press, Oxford (1965).
LENARD, P., *Great Men of Science,* 83-111, Bell & Sons, London (1933).
MANUEL, F. E., *A Portrait of Isaac Newton,* Harvard Univ. Press, Cambridge, Mass. (1968).
MORE, L. T., *Isaac Newton: a Biography,* Scribners, New York (1934).
RATTANSI, P. M., 'Isaac Newton and Gravity', *Pioneers of Science and Discovery,* Priory Press, London.
ROUSE, H. AND INCE, S., *History of Hydraulics,* Dover Publn. Inc., New York (1963).
WILLIAMS, T. I. (ed), *A Biographical Dictionary of Scientists,* 390-2, A & C Black, London (1969).

NUSSELT

KLING, G., Article on Nusselt, *Chemie Ingenieur Technik,* **24,** 597-608 (1952).
LÜCK, G., Article in *Gesunheit Ingenieur,* **74,** 7-8 (1953).
McADAMS, W. H., *Heat Transmission,* 95, McGraw-Hill, London (1951).
NUSSELT, E. K. W., Numerous Articles in *Zeit. Verein. deutsch. Ingr* (1900-1929).
NUSSELT, E. K. W., *Tech. Mech. Thermodynam.* **1,** 277, 417 (1930).

OHM

ASIMOV, I., *Asimov's Biographical Encyclopedia of Science and Technology,* 271-2, Pan Books, London (1975).
HART, I. B., *Makers of Science,* OUP, London (1930).
LENARD, P., *Great Men of Science,* 236-40, G. Bell & Sons, London (1933).
LOMMEL, E. (ed), *G. S. Ohm, Gesammelte Abhandlungen* (1892).
MAGIE, W. F., *A Source Book in Physics,* 465, Harvard U.P., Cambridge, Mass. (1965).
WILLIAMS, T. I. (ed), *A Biographical Dictionary of Scientists,* 397, A & C Black, London (1969).

OTTO

ASIMOV, I., *Asimov's Biographical Encyclopedia of Science and Technology,* 401, Pan Books, London (1975).
'Beau de Rochas, Otto and the four-stroke cycle engine', *The Engineer,* **170,** 5-6 (1940).
BRYANT, L., 'The Silent Otto,' *Technology and Culture,* **7,** 184-200 (1966).
CLERK, D., *Gas, Petrol and Oil Engine,* vol 1, Longmans Green, London (1910).
CROSSLEY, F. W., 'Otto and Langen's atmospheric gas engine, and some other gas engines, *Proc. Mech. E.,* **25,** 191-216 (1875).
GREAVES, W. F. AND CARPENTER, J. H., *A Short History of Mechanical Engineering,* 81-2, Longmans, London (1969).
HELDT, P. M., 'Origin of the four-stroke cycle,' *Automotive Industries,* **80,** 192-7 (1939).
LANGEN, A., *Niclaus August Otto, der Schöpfer des Verbrennungsmotors,* Stuttgart (1949).
MATSCHOSS, C., *Great Engineers,* 280-92, G. Bell & Sons, London (1939).
NAGEL, A., 'Die Bedeutung Ottos und Langens für die Entwicklung des Verbrennungsmotors', *Zeitschrift des Vereines deutscher Ingenieure,* **75,** 827 (1931), **80,** 1289 (1936).
WILLIAMS, T. I. (ed), *A Biographical Dictionary of Scientists,* 400, A & C Black, London (1969).
WILLIAMS, T. I. (ed), *A History of Technology,* **5,** 158, Clarendon Press, Oxford (1978).

PARSONS

APPLEYARD, R., *Charles Parsons; His Life and Work,* Constable, London (1933).
BURSTALL, A. F., *A History of Mechanical Engineering,* 339-41, 356-9, 381, 390, Faber & Faber, London (1963).
GLAZEBROOK, SIR R., Obituary Notice, *Nature,* **137,** 314 (1931).
Hon., *Sir Charles Algernon Parsons, Scientific Papers and Addresses,* ed. Hon. G. L. Parsons (inc. memoir by Lord Rayleigh), CUP, Cambridge (1934).
MATSCHOSS, C., *Great Engineers,* 304-14, G. Bell & Sons, London (1939).
PARSONS, R. H., *The Development of the Parsons' Steam Turbine,* Constable, London (1936).
WILLIAMS, T. I. (ed), *A Biographical Dictionary of Scientists,* 403-4, A & C Black, London (1969).

PASCAL

BALL, W. W. R., *A Short Account of the History of Mathematics,* Dover Publns. Inc., New York (1908).
BURSTALL, A. F., *A History of Mechanical Engineering,* 170-1, 252, Faber & Faber, London (1963).
GREAVES, W. F. AND CARPENTER, J. H., *A Short History of Mechanical Engineering,* 57, 77, Longmans, London (1969).
LENARD, P., *Great Men of Science,* 48-50, G. Bell & Sons, London (1933).
MESNARD, J., *Pascal, his Life and Works,* Philosophical Library, New York (1952).
MOULTON, F. R. AND SCHIFFERES, J. J., *The Autobiography of Science,* 145, John Murray, London (1963).
ROUSE, H. AND INCE, S., *History of Hydraulics,* 45, 76-9, 87, 90, 103, 127, Dover Publns. Inc., New York (1963).
WILLIAMS, T. I. (ed), *A Biographical Dictionary of Scientists,* 405-6, A & C Black, London (1969).

PELTON

BARNA, P. S., *Fluid Mechanics for Engineers,* 349, Butterworths, London (1957).
BREED, E. M., 'The Pelton Wheel', *Mech. Engng.,* **52,** 386-9 (1930).
BURSTALL, A. F., *A History of Mechanical Engineering,* 360, 406, Faber & Faber, London (1963).

DURAND, W. F., 'The Pelton Wheel', *Mech Engng.*, **61**, 447-54, 511-18 (1939).
Encyclopaedia Britannica, Micropaedia, **7**, Knowledge in Depth, 18.
KAEMPFFERT, W. (ed), *A Popular History of American Invention*, 533, 536, Charles Scribner's Sons, New York (1924).
ROUSE, H. AND INCE, S., *History of Hydraulics*, 187-8, Dover Publns. Inc., New York (1963).

PITOT

BARNA, P. S., *Fluid Mechanics for Engineers*, 110, 113, Butterworths, London (1957).
BURSTALL, A. F., *A History of Mechanical Engineering*, 172, Faber & Faber, London (1963).
FOLSON, R. G., 'Review of the Pitot tube', *Trans. Am. Soc. Med. Engrs.*
LINFORD, A., *Flow Measurements and Meters*, E. & F. Spon, London (1949).
PITOT, H., 'Description, 'd'une machine pour mesurer la vitesse des eaux et le sillage des vaisseaux', *Histoire de l'Academie des Sciences* (1732).
ROUSE, H. AND INCE, S., *History of Hydraulics*, 114-16, 134, 138, 175, 232, Dover Pubns. Inc., New York (1963).

PLANCK

DUGAS, R. (trans Maddox, J.R.), *A History of Mechanics*, 549-50, Editions du Griffon, Neuchâtel, Switzerland (1955).
GILLISPIE, C. C. (ed), *Dictionary of Scientific Biography*, **11**, 7-17, Charles Scribner's Sons, New York (1971).
HURD, D. L. AND KIPLING, J. J., *The Origins and Growth of Physical Science*, 2 vols, 402, Penguin Books (1964).
MOULTON, F. R. AND SCHIFFERES, J. J., *The Autobiography of Science*, 536, John Murray, London (1963).
Obituary Notices of Fellows of the Royal Society (1948).
PLANCK, M., *Autobiography* (English translation Gayner, F.), Williams & Norgate, London (1950).
WILLIAMS, T. I., *A Biographical Dictionary of Scientists*, 418-420, A & C Black, London (1969).

POISEUILLE

BURSTALL, A. F., *A History of Mechanical Engineering*, 322, Faber & Faber, London (1963).
GILLISPIE, C. C. (ed), *Dictionary of Scientific Biography*, **11**, 62-4, Charles Scribner's Sons, New York (1971).
Grande Larousse Encyclopédique, **8**, Libraire Larousse, Paris.
HAGEN, G. H. L., 'Ueber die Bewegung des Wassers in engen cylindrischen Röhren'; Poggendorf (1839).
ROUSE, H. AND INCE, S., *History of Hydraulics*, 160, 161, 164, Dover Publns. Inc., New York (1963).
SCHILLER, L., *Drei Klassiker der Strömungslehre: Hagen, Poiseuille, Hagenbach*, Akademische Verlagsgesellschaft, Leipzig (1933).

POISSON

BURSTALL, A. F., *A History of Mechanical Engineering*, 210, Faber & Faber, London (1963).
Grande Larousse Encyclopédique, **8**, Libraire Larousse, Paris.
POISSON, S. D., 'Memoire sur les lois du mouvement des fluides', *Mem. de l'Acad. Roy. Sci.*, **6** (1827).
ROUSE, H. AND INCE, S., *History of Hydraulics*, 195, 196, 218, Dover Publns. Inc., New York (1963).
TIMOSHENKO, S. P., *History of Strength of Materials*, 111-14, McGraw-Hill, New York (1953).
WILLIAMS, T. I. (ed), *A Biographical Dictionary of Scientists*, 423, A & C Black, London (1969).

PRANDTL

BARNA, P. S., *Fluid Mechanics for Engineers*, 177, 198, Butterworths, London (1957).
BURSTALL, A. F., *A History of Mechanical Engineering*, 360, 397, Faber & Faber, London (1963).
GOLDSTEIN, S. (ed), *Modern Developments in Fluid Mechanics*, OUP, Oxford (1938).
PRANDTL, L., *The Physics of Solids and Fluids*, 2 vols, 2nd edn, Blackie, London (1936).
PRANDTL, L. AND TIETJENS, O. G., *Applied Hydro- and Aero-mechanisms*, McGraw-Hill, London (1934).
PRANDTL, L., 'Göttingen wind tunnel for testing aircraft models', *NACA*, TN66 (Nov. 1920).
ROUSE, H. AND INCE, S., *History of Hydraulics*, 229-35, 237-39, 242, Dover Publns. Inc., New York (1963).
TIMOSHENKO, S. P., *History of Strength of Materials*, 392-7, McGraw-Hill, New York (1953).

PYTHAGORAS

BALL, W. W. R., *A Short Account of the History of Mathematics*, Dover Publns. Inc., New York (1908).
BURSTALL, A. F., *A History of Mechanical Engineering*, 66, Faber & Faber, London (1963).
CLAGETT, M., *Greek Science in Antiquity*, Bilbo & Tannen, New York (1955).
GORMAN, P., *Pythagoras, A Life*, Routledge & Kegan Paul, London (1979).
GUTHRIE, W. K. C., *A History of Greek Philosophy*, vol 1, CUP, Cambridge (1962).
LENARD, P., *Great Men of Science*, 1, 5, G. Bell & Sons, London (1933).
ROUSE, H. AND INCE, S., *History of Hydraulics*, 11, Dover Publns. Inc., New York (1963).
TURNBULL, H. W., *The Great Mathematicians*, 1, Methuen, London (1929).
WILLIAMS, T. I. (ed), *A Biographical Dictionary of Scientists*, 430, A & C Black, London (1969).

RANKINE

BURSTALL, A. F., *A History of Mechanical Engineering*, 212, 278, 286, 353, 368, Faber & Faber, London (1963).
GARDINER, A. W., 'Automotive steam power — 1973', Paper 730617 (inc. bibliography on Rankine cycle powerplants), *Soc. Automotive Engrs.*
GILLISPIE, C. C. (ed), *Dictionary of Scientific Biography*, **11**, 291-5, Charles Scribner's Sons, New York (1971).
RANKINE, W. J. M., *A Manual of Applied Mechanics*, Charles Griffin & Co, London (1858).
RANKINE, W. J. M., *A Manual of the Steam Engine and Other Prime Movers'*, Charles Griffin & Co, London (1859).
TIMOSHENKO, S. P., *History of Strength of Materials*, 197-202, McGraw-Hill, New York (1953).
WILLIAMS, T. I. (ed), *A Biographical Dictionary of Scientists*, 434, A & C Black, London (1969).

RAYLEIGH

BURSTALL, A. F., *A History of Mechanical Engineering*, 288, Faber & Faber, London (1963).
CAMERON, A., *Principles of Lubrication*, Longmans, London (1966).
Les Prix Nobel en 1904, 1907.
LINDSAY, R. B., *Lord Rayleigh, The Man & His Work*, Pergamon, Oxford (1970).
LORD RAYLEIGH, 'Notes on the theory of lubrication', *Phil. Mag.*, **35**, 1-12 (1918).
ROUSE, H. AND INCE, S., *History of Hydraulics*, 211-13, Dover Publns. Inc., New York (1963).
STRUTT, R. J., *Life of John Williams Strutt, Third Baron Rayleigh*, Univ. of Wisconsin Press, London (1968).

TEMPLE, G. AND BICKLEY, W. G., *Rayleigh's Principle and its Applications to Engineering: the Theory and Practice of the Energy Method for the Approximate Determination of Critical Loads and Speeds,* OUP, London (1933).
TIMOSHENKO, S P., *History of Strength of Materials,* 334-9, McGraw-Hill, New York (1953).
WILLIAMS, T. I. (ed), *A Biographical Dictionary of Scientists,* 435, A & C Black, London (1969).

REYNOLDS

Beauchamp Tower. Five Reports on the Research Committee on Friction, *Proc. I. Mech. E.* (1883-91).
BURSTALL, A. F., *A History of Mechanical Engineering,* 286, 317, 321, 323, 355, 360, Faber & Faber, London (1963).
GIBSON, A. H., *Osborne Reynolds,* Longmans Green, London (1946).
McDOWELL. D. AND JACKSON, J. D., *Osborne Reynolds and Engineering Science Today,* Manchester Univ. Press, Manchester (1970).
REYNOLDS, O., 'An experimental investigation of the circumstances which determine whether the motion of water shall be direct or sinuous and of the law of resistance in parallel channels, *Phil. Trans. Roy. Soc.,* **174** (1883).
REYNOLDS, O., *Scientific Papers,* 2 vols, CUP, Cambridge (1900-1).
ROUSE, H. AND INCE, S., *History of Hydraulics,* 182, 191, 206-12, 218, 224, 238-9, Dover Publns. Inc., New York (1963).
WILLIAMS, T. I. (ed), *A Biographical Dictionary of Scientists,* 440-1, A & C Black, London (1969).

RUMFORD

ALDEN THOMPSON, J., *Count Rumford of Massachusset's,* Farrar & Rinehart, New York (1935).
BROWN, S. C., *Count Rumford, Physicist Extraordinary,* Heinemann, London (1962).
BROWN, S. C., *Benjamin Thompson — Count Rumford on the Nature of Heat,* Pergamon Press, London (1967).
LARSEN, E., *An American in Europe,* Rider & Co, New York (1953).
LENARD, P., *Great Men of Science,* 170-5, Bell & Sons, London (1933).
MATSCHOSS, C., *Great Engineers,* 128-30, Bell & Sons, London (1939).
WILLIAMS, T. I., *A Biographical Dictionary of Scientists,* 452-3, A & C Black, London (1969).
WILSON, M., 'Count Rumford', *Scientific American,* **203,** no 4, 158-68 (1960).

SIEMENS

BURSTALL, A. F., *A History of Mechanical Engineering,* 291, 293, Faber & Faber, London (1963).
DUNSHEATH, P., *A History of Electrical Engineering,* Faber & Faber, London (1962).
GILLISPIE, C. C. (ed), *Dictionary of Scientific Biography,* Chas. Scribner's Sons, New York (1971).
GREAVES, W. F. AND CARPENTER, J. H., *A Short History of Mechanical Engineering,* 86, 87, 113, Longmans, London (1969).
LENARD, P., *Great Men of Science,* 255, 268, 299, 342, Bell & Sons, London (1933).
MATSCHOSS, C., *Great Engineers,* 261-80, Bell & Sons, London (1939).
Recollections of Werner von Siemens, new edn., Land Humphries, London (1966).
WILLIAMS, T. I. (ed), *A Biographical Dictionary of Scientists,* 474-5, A & C Black, London (1969).

STEPHENSON

BURSTALL, A. F., *A History of Mechanical Engineering,* 203, 267-269, 273, Faber & Faber, London (1963).
MacHAY, C., *George Stephenson & the Project of Railway Enterprise,* Andrew Reid, Newcastle (1881).
MATSCHOSS, C., *Great Engineers,* 171-189, Bell & Sons, London (1939).
ROLT, L. T. C., *George and Robert Stephenson,* Longmans Green (1960).
SKEAT, W. O., *George Stephenson,* Institution of Mechanical Engineers, London (1973).
SMILES, S., *Life of George Stephenson,* John Murray, London (1864, 1904 edn).
THOMAS, J., *The Story of George Stephenson,* OUP, Oxford (1952).
WILLIAMS, T. I. (ed), *A Biographical Dictionary of Scientists,* 493-4, A & C Black, London (1969).

STIRLING

BURSTALL, A. F., *A History of Mechanical Engineering,* 274-6, 350, 437, Faber & Faber, London (1963).
GREAVES, W. F. AND CARPENTER, J. H., *A Short History of Mechanical Engineering,* 80, 114, Longmans, London (1969).
KOLIN, I., *The Evolution of the Heat Engine, Thermodynamic Atlas 2,* Longmans, London (1972).
PHILIPS, N. V., Gloeilampenfabrieken. Patent No. 605922 (1946). Improvements in or relating to hot gas reciprocating engines.
STIRLING R., Patent No. 4081 (1816), Diminishing the Consumption of Fuel; Engine capable of being applied to the Moving of Machinery.
WALKER, G., 'Operations cycle of the Stirling engine with particular reference to the function of the regenerator', *J. Mech. Engng. Sci.,* **3,** 394-408 (1961).
WALKER, G., Stirling-cycle Machines (bibliography pp. 145-52), Clarendon Press, Oxford (1973).

STOKES

BURSTALL, A. F., *A History of Mechanical Engineering,* 287, Faber & Faber, London (1963).
LENARD, P., *Great Men of Science,* 101, 369, Bell & Sons, London (1933).
MAGIE, W. F., *A Source Book in Physics,* Harvard Univ. Press, Cambridge, Mass. (1935).
Proc. Roy. Soc., **75,** 199 (1905).
ROUSE, H. AND INCE, S., *History of Hydraulics,* 197-218, Dover Publns. Inc., New York (1963).
STOKES, G. G., On the theories of the internal friction of fluids in motion and of the equilibrium and motion of elastic bodies, *Trans. Cambridge Phil. Soc.,* **8** (1845).
STOKES, G. G., *Memoirs and Scientific Correspondence* (1907), reprinted Johnson Reprint, New York (1971).
TIMOSHENKO, S. P., *History of Strength of Materials,* 225-9, McGraw-Hill, New York (1953).
WILLIAMS, T. I. (ed), *A Biographical Dictionary of Scientists,* 496, A & C Black, London (1969).

TESLA

BECKHARD, A. J., *Nikola Tesla, Electrical Genius,* Dennis Dobson, London (1961).
DUNSHEATH, P., *A History of Electrical Engineering,* Faber, London (1962).
Illustrated London News, 117, 335 (1900).
J.I.E.E., **90,** 457 (1943).
KAEMPFFERT, W. (ed), *A Popular History of American Invention,* 526, Charles Scribner's Sons, New York and London (1924).
MARTIN, T. C., 'The Inventions, Researches and Writings of Nikola Tesla', *The Electrical Engineer,* London (1894).
O'NEILL, J. J., *Prodigal Genius,* Ives Washburn Inc., New York (1944), Spearman, London (1968), *Trans. Am. Inst. Elec. Engrs.,* **62,** 76, (1943).
Tesla Society. Bibliography; Dr. Nikola Tesla, Mineapolis (1954).
TESLA, N., *Experiments with Alternating Currents of High Potential and Frequency,* McGraw-Hill, New York (1904).
WILLIAMS, T. I. (ed), *A Biographical Dictionary of Scientists,* 507, A & C Black, London (1969).

TIMOSHENKO
BURSTALL, A. F., *A History of Mechanical Engineering,* 290, Faber & Faber, London (1963).
Collected Works of Stephen P. Timoshenko, (with biog.), McGraw-Hill, London (1953).
TIMOSHENKO, S. P., *History of Strength of Materials,* McGraw-Hill, New York (1953).
TIMOSHENKO, S. P., *As I Remember,* D. Van Nostrand, Princeton (1968) (Autobiography).
TIMOSHENKO, S. P., *Engineering Education in Russia,* McGraw-Hill, New York (1959).

TORICELLI
BURSTALL, A. F., *A History of Mechanical Engineering,* 171, Faber & Faber, London (1963).
CALDONAZZO, B., CARRUCCIO, E. AND RONCHI, V., *E. Toricelli,* Casa Editricé Universiterra, Florence (1951).
GREAVES, W. F. AND CARPENTER J. H., *A Short History of Mechanical Engineering,* Longmans, London (1969).
LENARD, P., *Great Men of Science,* 48-50, Bell & Sons, London (1933).
MAGIE, W. F., *A Source Book in Physics,* 111, Harvard Univ. Press. CUP, Cambridge, Mass. (1965).
ROUSE, H. AND INCE, S., *History of Hydraulics,* 61-9, 76-80, 89, Dover Publns. Inc., New York (1963).
TOKATY, G. A., *A History and Philosophy of Fluid Mechanics,* 56, G. T. Foulis & Co, Henley-on-Thames (1971).
WILLIAMS, T. I. (ed), *A Biographical Dictionary of Scientists,* 516-7, A & C Black, London (1969).

TREVETHICK
BARTON, D. B., *The Cornish Beam Engine: a survey of its history and development in the mines of Cornwall and Devon from before 1800 to the present day,* Barton, Truro (1965).
DICKINSON, H. W., *A Short History of the Steam Engine,* CUP, Cambridge (1938).
DICKINSON, H. W. AND TITELY, A., *Richard Trevethick, the Engineer and the Man,* CUP, Cambridge (1934).
HODGE, J., *An Illustrated Life of Richard Trevethick,* 1771-1833, Shire Pubns, Aylesbury (1973).
MATSCHOSS, C., Great Engineers, 154-171, Bell & Sons, London (1939).
PENDRED, L. ST. L., 'Trevethick', *Trans. Newcomen Soc.,* **1**, 34 (1920).
WILLIAMS, T. I. (ed), *A Biographical Dictionary of Scientists,* 519-20, A & C Black, London (1969).

VENTURI and HERSCHEL
BARNA, P. S., *Fluid Mechanics for Engineers,* 104, Butterworths, London (1957).
BURSTALL, A. F., *A History of Mechanical Engineering,* 253, Faber & Faber, London (1963).
'Clemens Herschel's invention of the Venturi meter', *Engineering,* **140**, 110-111 (1935).
HERSCHEL, C., 'The Venturi Meter; an instrument making use of a new method of gauging water; applicable to the cases of very large tubes and of small value only of the liquid to be gauged', *Trans. Am. Soc. Civ. Engrs,* **17**, 228-58 (1887).
'Herschel Will Cites Need for Hydraulic Research', *Engineering News Record,* **105**, No 1 (1930).
LINDLEY, D., 'An experimental investigation of the flow in a classical venturi meter, *Proc. I. Mech. E.,* 1969-70, **184**, 133-160 (1969-70).
ROUSE, H. AND INCE, S., *History of Hydraulics,* 134-7, 189-90, 226, Dover Publns. Inc., New York (1963).

VOLTA
Grande Larousse Encyclopédique, **10**.
DIBNER, B., *Allesandro Volta and the Electric Battery,* Franklin Watts, New York (1964).
DUNSHEATH, P., *A History of Electrical Engineering,* 29-31, Faber & Faber, London (1962).
LENARD, P., *Great Men of Science,* 158-170, Bell & Sons, London (1933).
OLBY, R. C. (ed), *Late 18th Century European Scientists,* 127, Pergamon, London (1966).
WILLIAMS, T. I. (ed), *A Biographical Dictionary of Scientists,* 535, A & C Black, London (1969).

WATT
BURSTALL, A. F., *A History of Mechanical Engineering,* Faber & Faber, London (1963).
CROWTHER, J. G., *Scientists of the Industrial Revolution,* Cresset Press, London (1962).
DICKINSON, H. W., *James Watt and the Steam Engine,* OUP, Oxford (1927).
DICKINSON, H. W., *James Watt,* CUP, Cambridge (1936).
DICKINSON, H. W., *James Watt, Craftsman and Engineer,* David & Charles, Newton Abbot (1967).
LENARD, P., *Great Men of Science,* 130-6, Bell & Sons, London (1933).
MATSCHOSS, C., Great Engineers, 95-110, Bell & Sons, London (1939).
MUIRHEAD, J. P., *The Origin and the Progress of the Mechanical Inventions of James Watt,* illustrated by his correspondence with his friends and the specifications of his patents, 3 vols, Murray, London (1854).
ROLT, L. T. C., *James Watt,* Longmans Green, London (1962).
WILLIAMS, T. I. (ed), *A Biographical Dictionary of Scientists,* 544-5, A & C Black, London (1969).

WEBER
ASIMOV, I., *Asimov's Biographical Encyclopaedia of Science and Technology,* 316, Pan Books, London (1975).
GILLISPIE, C. C. (ed), *Dictionary of Scientific Biography,* **14**, 203-9, Charles Scribner's Sons, New York (1971).
LENARD, P., *Great Men of Science,* 263-70, Bell & Sons, London (1933).
Nature, **44**, 206, 229, 272 (1891).
ROUSE, H. AND INCE, S., *History of Hydraulics,* 144-6, 199, Dover Publns. Inc., New York (1963).
WILLIAMS, T. I. (ed), *A Biographical Dictionary of Scientists,* 545, A & C Black, London (1969).

WHITWORTH
BRADLEY, I., *A History of Machine Tools,* (Biog.) Model & Allied Pubns, Hemel Hempstead (1972).
British Standard 84:1956. *Parallel screw threads of Whitworth form,* British Standards Institution, London.
BURSTALL, A. F., *A History of Mechanical Engineering,* 214, 225, 281, 318, Faber & Faber, London (1963).
GREAVES, W. F. AND CARPENTER, J. H., *A Short History of Mechanical Engineering,* 119-121, Longmans, London (1969).
MUSSON, A. E., 'Sir Joseph Whitworth', *The Vickers Magazine, II* (Summer 1966).
WHITWORTH, J., 'A Uniform System of Screw Threads', Paper at Inst. Civ. Engrs, **1**, 157-60 (1841).
WHITWORTH, J., Address to the Inst. Mech. Engrs. at Glasgow (1856).
WILLIAMS, T. I. (ed), *A Biographical Dictionary of Scientists,* 556-7, A & C Black, London (1969).

YOUNG

ASIMOV, I., *Asimov's Biographical Encyclopaedia of Science and Technology*, 241-2, Pan Books, London (1975).
GILLISPIE, C. C. (ed), *Dictionary of Scientific Biography*, **14**, 562-72, Charles Scribner's Sons, New York (1971).
LENARD, P., *Great Men of Science*, 196-8, Bell & Sons, London (1933).
MAGIE, W. F., *A Source Book in Physics*, 95, Harvard Univ. Press, Cambridge, Mass. (1965).
OLBY, R. C. (ed), *Early 19th Century European Scientists*, 67, Pergamon (1967).
TIMOSHENKO, S. P., *History of Strength of Materials*, 90-8, McGraw-Hill, New York (1953).
WILLIAMS, T. I. (ed), *A Biographical Dictionary of Scientists*, 572, A & C Black, London (1969).
WOOD, A., *Thomas Young: Natural Philosopher*, CUP, Cambridge (1954).